Innovations and Implications in Molecular Diagnostics

This book introduces the field of molecular diagnostics, tracing its evolution from the earliest days of genetic research to cutting-edge advancements. It begins with a historical overview, setting the stage for an understanding of the foundational principles and techniques that underpin current practices in DNA analysis. It also provides the intricate details of various molecular techniques, such as advanced polymerase chain reaction (PCR) methods and next-generation sequencing, and their impact in the field of genetics and diagnostics. The book also explores the significant role of bioinformatics as the backbone of modern diagnostics and examines the changes brought about by high-throughput technologies and nanotechnology in molecular research and diagnostics. The importance of molecular imaging techniques in visualizing biological processes is discussed, along with the advancements in RNA analysis and their diagnostic implications. The book addresses the emerging fields of metabolomics and proteomics in molecular diagnostics and the transformative impact of artificial intelligence in this field. The book concludes with a thoughtful discussion on the ethical and regulatory considerations in molecular diagnostics. This book is intended for researchers and students of the fields of molecular diagnostics, biotechnology, and genomics.

Key Features:

- Provides historical overview and recent advancements in the field of molecular diagnostics
- Explores foundational principles and advanced techniques in DNA analysis, including detailed insights into PCR methods
- Covers the applications and impact of next-generation sequencing in genetics and diagnostics
- Discusses the revolutionary influence of high-throughput technologies and nanotechnology in molecular research and diagnostics
- Examines the roles of metabolomics, proteomics, and artificial intelligence in molecular diagnostics

Innovations and Implications in Molecular Diagnostics

Edited by
Qamre Alam, Misbahuddin Rafeeq,
and Muhammad Umair

CRC Press
Taylor & Francis Group
Boca Raton London New York

CRC Press is an imprint of the
Taylor & Francis Group, an **informa** business

Designed cover image: Shutterstock

First edition published 2026
by CRC Press
2385 NW Executive Center Drive, Suite 320, Boca Raton FL 33431

and by CRC Press
4 Park Square, Milton Park, Abingdon, Oxon, OX14 4RN

CRC Press is an imprint of Taylor & Francis Group, LLC

ISBN: 978-1-032-78725-1 (hbk)
ISBN: 978-1-032-78726-8 (pbk)
ISBN: 978-1-003-48917-7 (ebk)

DOI: 10.1201/9781003489177

Typeset in Sabon
by KnowledgeWorks Global Ltd.

Contents

About the editors

Dr. Qamre Alam is currently serving as a Medical Research Scientist and Laboratory Supervisor at the Molecular Genomics and Precision Medicine Department of ExpressMed Diagnostics and Research, Zinj, Kingdom of Bahrain. Previously, from 2019 to 2022, he held the position of Medical Research Scientist at the Medical Genomics Research Department at King Abdullah International Medical Research Center, part of King Saud bin Abdulaziz University for Health Science, Ministry of National Guard–Health Affairs, Riyadh, Saudi Arabia. Prior to that, he worked as a lecturer from 2011 to 2019 at the King Fahd Medical Research Center, King Abdulaziz University, Jeddah, Saudi Arabia. He earned his MSc from Jamia Hamdard University, New Delhi, India, and his PhD in Biotechnology from JJTU, Rajasthan, India. He has an impressive publication record, with over 75 research articles in the fields of human genetics and cancer biology. Dr. Alam is a molecular geneticist with a strong research interest in rare genetic disorders. He has actively contributed to various projects related to rare genetic disorders and preventive genomic medicine, including non-invasive prenatal testing and preimplantation genetic screening. Additionally, he serves as an Associate Editor and reviewer for several international peer-reviewed journals of high repute.

Dr. Misbahuddin Rafeeq is a distinguished physician, accomplished scientist, and dedicated medical educator. He earned his MBBS and MD degrees in 2009 from Aligarh Muslim University, Aligarh, India. Currently, he holds the position of Associate Professor at the Faculty of Medicine (Rabigh) within King Abdulaziz University, Jeddah, Saudi Arabia. Dr. Rafeeq is an esteemed member of the editorial boards of several reputable journals. His wealth of expertise has been showcased through his active participation in numerous scientific events spanning across regions such as India, Southeast Asia, Asia-Pacific, Europe, the United Kingdom, the Middle East, Africa, the Balkans, and the United States. His outstanding contributions have earned him several prestigious awards, including the Excellent Leadership Award from the Islamic Development Bank in Saudi Arabia, the ESP fellowship from NIHES in the Netherlands, the Distinguished Achievement Award from KAU in Saudi Arabia, and the Bronze Prize from the Diabetic Congress in the United States. Dr. Rafeeq is affiliated with esteemed organizations and scientific societies, including RCPEdin and the Faculty of Pharmaceutical Medicine at RCP London, ACCP and ASPET in the United States, ISoP, and the Royal Society of Medicine in the United Kingdom. Additionally, he serves as an honorary consultant on various educational boards and holds the position of International Secretary for SoPI in India. Dr. Rafeeq boasts an impressive publication record with over 60 high-impact and indexed publications to his credit. Furthermore, he has secured funding for several large grants, where he has served as both a Principal and Co-investigator. His research primarily focuses on the identification of novel genetic mutations in hereditary syndromes, the exploration of novel therapeutic targets and proteins through molecular docking, PK/PD, toxicity studies, the study of the effects of natural compounds on animal models, and the broader field of lipid pharmacogenomics.

Dr. Muhammad Umair currently holds the position of Associate Research Scientist and serves as the Team Leader of Functional Studies at the Medical Genomics Research Department within the King Abdullah International Medical Research Center, part of King Saud bin Abdulaziz University for Health Science, Ministry of National Guard–Health Affairs in Riyadh, Saudi Arabia. Additionally, he serves as a "Research Advisor" at the Department of Life Sciences within the School of Science at the University of Management and Technology in Lahore, Pakistan. Dr. Umair is a molecular geneticist with a particular focus on rare genetic disorders. His research involves multiple projects dedicated to rare genetic disorders and therapeutic genetics. He is currently engaged in research initiatives such as the Genetics and Rare Disease Program, Genetic and Functional Characterization of Rare Skeletal Disorders, Genodermatosis, Preventative Genome Medicine for Inherited Genetic Disorders, Functional Characterization of Genes Associated with Neurodevelopmental Disorders, Genetic and Rare Disease Registry, and Therapeutic Genomics. Dr. Umair earned his MSc, MPhil, and PhD in Biochemistry/Molecular Biology, specializing in Human Molecular Genetics, from Quaid-i-Azam University, Islamabad, Pakistan. His extensive contributions to the field are reflected in his publication record, which comprises over 140 research articles in the domain of human genetics. He also actively participates as an Associate Editor and reviewer for several international peer-reviewed journals. In recognition of his outstanding achievements in genetics, Dr. Umair was honored with the 2nd Dr. Sajjad Aslam Shami Gold Medal in Genetics by the Applied Zoological Society of Pakistan in March 2021.

Chapter 1

Evolution of molecular diagnostics

Genetic foundations to modern technological innovations

*Ibrahim A. Almazni**

Department of Clinical Laboratory Sciences, College of Applied Medical Sciences, Najran University, Najran, Saudi Arabia

Basmah Massad Alenzi

Administration of Laboratory and Blood Bank Tabuk Health Cluster, Tabuk, Saudi Arabia

Doha A. Albalawi

Department of Biology, Biodiversity Genomics Unit, Faculty of Science, University of Tabuk, Tabuk, Saudi Arabia

Aishah Alatawi

Department of Biology, Faculty of Science, University of Tabuk, Tabuk, Saudi Arabia

HISTORICAL PERSPECTIVE

The field of molecular diagnostics has undergone profound transformations from its rudimentary beginnings to its current sophisticated state. This journey encompasses significant milestones that span centuries, reflecting the cumulative progress in understanding genetic material and developing technologies to analyze it (**Table 1.1**).

The concept of genetic inheritance can be traced back to ancient times. Hippocrates, widely recognized as the father of medicine, contemplated the mechanism by which qualities are passed on from parents to their progeny. Nevertheless, it was only during the mid-19th century that Gregor Mendel conducted his studies, leading to the scientific establishment of the laws of genetic inheritance. Mendel established the presence of distinct genetic units, subsequently referred to as genes, that follow particular patterns of inheritance through his work with pea plants, which was published in 1866 (Alkhilaiwi & Khan, 2022). Significant progress was made in the late 19th century with the discovery of nucleic acids. In 1869, Friedrich Miescher isolated a substance from the nuclei of white blood cells, which he named "nuclein." Subsequently, the substance was identified as DNA. The discovery and isolation of nuclein marked the beginnings of molecular biology, but its importance was not immediately recognized (Mazzarello, 1999). The early 20th century was marked by pivotal discoveries that laid the groundwork for molecular diagnostics. In 1910, Thomas Hunt Morgan's work with fruit flies provided evidence that genes are located on chromosomes, solidifying the chromosome theory of inheritance. This era also saw the discovery of the transforming principle by Avery, MacLeod, and McCarty (1944),

* Corresponding author.

DOI: 10.1201/9781003489177-1

Table 1.1 Key milestones in the evolution of molecular diagnostics

Year	*Milestone*	*Contributor(s)*	*Description and impact*
1866	Principles of genetic inheritance	Gregor Mendel	Mendel's experiments with pea plants established the basic principles of heredity, introducing the concept of discrete hereditary units, later termed genes.
1953	Discovery of DNA double helix structure	James Watson, Francis Crick	The double helix model explained the molecular basis of genetic replication and transcription, fundamentally changing genetics and molecular biology.
1973	Creation of the first genetically modified organism	Stanley Cohen, Herbert Boyer	Demonstrated the use of plasmids and restriction enzymes to insert DNA from one organism into another, paving the way for genetic engineering.
1983	Polymerase Chain Reaction (PCR)	Kary Mullis	PCR allowed for the amplification of specific DNA sequences, revolutionizing genetic testing, forensic analysis, and infectious disease diagnosis.
2003	Completion of the Human Genome Project	International Consortium	Provided a comprehensive reference for human genetic research, identifying thousands of genes and genetic markers, significantly advancing genomic medicine.

which identified DNA as the material responsible for genetic transformation in bacteria. In 1953, James Watson and Francis Crick made a significant discovery by revealing the double helix structure of DNA, which was determined using X-ray diffraction images obtained by Rosalind Franklin. This finding fundamentally changed the direction of molecular biology and genetics by providing a molecular basis for understanding transcription and genetic replication (Watson & Crick, 1953).

The advent of recombinant DNA technology

The era of recombinant DNA technology began in the 1970s. Scientists were able to manipulate genetic material in unprecedented ways as a result of the development of restriction enzymes, which function as molecular scissors to cut DNA at specific sequences. In 1973, Stanley Cohen and Herbert Boyer exhibited the successful integration of DNA from one organism into another by using plasmids and restriction enzymes, thus producing the initial genetically modified organism (Cohen et al., 1973).

This period also witnessed the introduction of DNA sequencing techniques. Frederick Sanger developed the chain-termination method, also known as Sanger sequencing, in 1977. This method became the gold standard for DNA sequencing for many years and played a crucial role in early genomic projects (Sanger, Nicklen, & Coulson, 1977).

Polymerase chain reaction

The polymerase chain reaction (PCR), invented by Kary Mullis in 1983, is considered one of the most revolutionary tools in molecular diagnostics. PCR enables the amplification of precise DNA sequences, hence facilitating the detection of minuscule quantities of genetic material (**Figure 1.1**). The utilization of this technology brought about a significant transformation in the field of molecular biology, as it allowed for a diverse array of applications, including genetic testing, forensic analysis, and infectious disease identification (Mullis & Faloona, 1987).

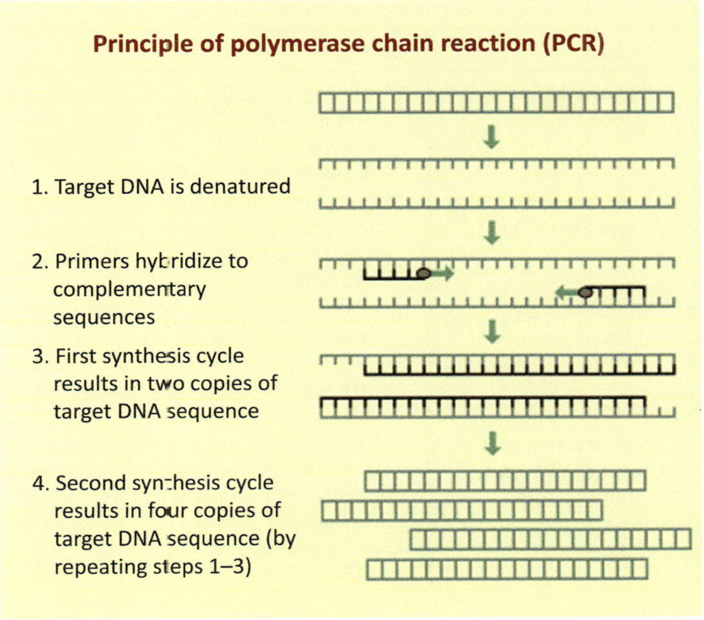

Figure 1.1 Schematic representation of the PCR process illustrating DNA denaturation, primer annealing, and extension for amplifying target DNA sequences.

The significance of PCR in clinical diagnosis cannot be overstated. The technology offered a quick, highly responsive, and precise approach to detect genetic alterations, identify infections, and investigate gene expression. The development of real-time PCR enabled the measurement of DNA and RNA, hence increasing its diagnostic capabilities (Heid et al., 1996).

The Human Genome Project

The Human Genome Project (HGP), initiated in 1990 and concluded in 2003, was a momentous global endeavor to determine the whole sequence of the human genome. The goal of the HGP, which was overseen by the National Institutes of Health and the US Department of Energy, was to map every gene in the human genome and comprehend its function. The initiative offered an extensive source of information for genetic research and diagnostics, identifying numerous genes and genetic markers linked to different disorders (Collins, Morgan, & Patrinos, 2003). The conclusion of the HGP marked the beginning of a new era in the field of genomic medicine. It supported the advancement of high-throughput sequencing technologies, such as next-generation sequencing (NGS), which allowed for the quick and cost-efficient sequencing of whole genomes and exomes. These advancements have had significant consequences for personalized medicine, enabling the detection of genetic predispositions to diseases and the creation of precise medicines (Mardis, 2008).

Next-generation sequencing

NGS technologies, which started in the mid-2000s, signify a substantial advancement in molecular diagnostics. NGS, in contrast to classical Sanger sequencing, enables the simultaneous sequencing of many DNA fragments, facilitating thorough examination of complete genomes or specific sections. The ability to process large amounts of data quickly has

Table 1.2 Applications of next-generation sequencing in molecular diagnostics

Application	*Benefits*	*Examples*
Cancer Genomics	Personalized treatment strategies	Identifying mutations in tumors such as BRAF in melanoma, EGFR in lung cancer, and HER2 in breast cancer; guiding the use of targeted therapies like vemurafenib, erlotinib, and trastuzumab, respectively.
Infectious diseases	Accurate pathogen identification	Sequencing of SARS-CoV-2 for tracking mutations and understanding transmission; identifying antibiotic resistance genes in bacterial infections to guide appropriate treatment.
Inherited disorders	Improved diagnosis of genetic disorders	Exome sequencing for identifying mutations in cystic fibrosis, Duchenne muscular dystrophy, and various neurodevelopmental disorders, leading to earlier and more accurate diagnoses and better-informed treatment plans.
Prenatal testing	Non-invasive detection of chromosomal abnormalities	NGS-based non-invasive prenatal testing for detecting Down syndrome, Edwards syndrome, and Patau syndrome by analyzing cell-free fetal DNA in maternal blood, offering safer alternatives to invasive procedures like amniocentesis.

transformed the field of diagnostics by offering comprehensive genetic information that may be utilized for disease identification, prediction, and therapy (**Table 1.2** and **Figure 1.2**) (Metzker, 2010).

Applications in molecular diagnostics

Cancer genomics

NGS has had a profound and transformational effect on the field of cancer genetics. NGS enables in-depth study of cancer genomes, facilitating the detection of mutations, insertions, deletions, and copy number changes that contribute to the growth and advancement of tumors. Precise genomic data is crucial for personalized medicine, as it facilitates the creation of customized treatments designed specifically for an individual's genetic makeup. Identifying precise genetic changes in a patient's tumor can provide valuable information

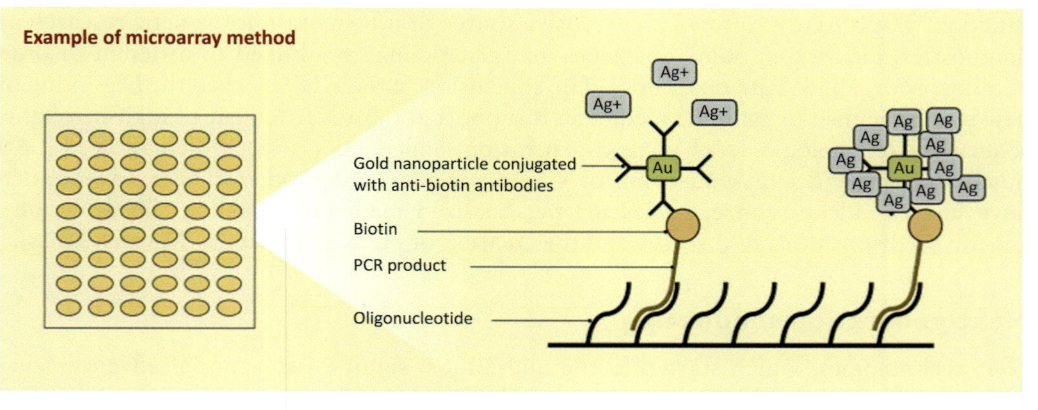

Figure 1.2 Overview of applications of next-generation sequencing in molecular diagnostics across various clinical areas.

for the selection of tailored treatments for diseases like melanoma, lung cancer, and breast cancer. This can greatly enhance the effectiveness of the treatment and improve the overall outcomes (Meyerson, Gabriel, & Getz, 2010).

Infectious diseases

NGS is a highly efficient method in the field of infectious diseases. It is used to accurately identify and characterize microorganisms, even in complicated clinical samples. This feature is especially beneficial for identifying newly developing or uncommon pathogens that might not be detected by conventional diagnostic techniques. Furthermore, NGS has the capability to detect antimicrobial resistance genes, helping clinicians make informed decisions regarding suitable treatments and enhancing their ability to effectively address antibiotic resistance. During the COVID-19 pandemic, NGS played a vital role in sequencing the SARS-CoV-2 virus, monitoring its genetic changes, and comprehending how it spreads (Gu et al., 2013).

Inherited genetic disorders

The use of NGS has greatly enhanced the accuracy and effectiveness of diagnosing inherited genetic illnesses. Exome sequencing, a technique that targets the protein-coding sections of the genome, enables the detection of mutations that cause monogenic illnesses. This method has enhanced the diagnostic output for illnesses such as cystic fibrosis, muscular dystrophy, and numerous neurodevelopmental abnormalities. NGS, with its extensive genomic analysis, allows for earlier and more precise diagnoses, which are crucial for the appropriate treatment and management of many conditions (Bamshad et al., 2011).

Prenatal testing

NGS has significantly improved non-invasive prenatal testing (NIPT). NGS-based NIPT examines the cell-free fetal DNA present in the mother's bloodstream to identify chromosomal abnormalities, including Down syndrome, Edwards syndrome, and Patau syndrome. This non-invasive method provides a safer option compared to conventional invasive treatments like amniocentesis. It decreases the likelihood of difficulties for both the mother and the fetus, while also delivering very precise outcomes (Bianchi et al., 2012).

Microbiome analysis

NGS has played a crucial role in enhancing our comprehension of the human microbiome, namely the gut microbiome. It also allows for a thorough analysis of the microbial communities present in the human body, offering valuable information about the impact of microbiome on both health and disease. Changes in the gut microbiota have been associated with illnesses such as inflammatory bowel disease, obesity, and mental health disorders, emphasizing the significance of maintaining a healthy microbiome for overall well-being. The comprehensive examination of microorganisms can result in innovative therapeutic techniques and individualized approaches to medicine (Qin et al., 2010).

The applications of NGS are vast and varied. It is used in oncology to identify genetic mutations in tumors, in infectious disease diagnostics to detect and characterize pathogens, and in genetic testing to diagnose hereditary disorders. NGS has also facilitated the study of complex diseases by allowing for the identification of rare genetic variants and their associations with disease phenotypes (Goodwin, McPherson, & McCombie, 2016).

CRISPR-Cas9 and genome editing

The discovery of the CRISPR-Cas9 system in the early 2010s marked another milestone in molecular biology. Originally identified as a bacterial immune system that uses RNA-guided endonucleases to cut foreign DNA, CRISPR-Cas9 was adapted for genome editing in eukaryotic cells. This technology allows for precise modifications of specific DNA sequences, making it a powerful tool for both research and clinical applications (Doudna & Charpentier, 2014).

In molecular diagnostics, CRISPR-based methods are being developed for detecting genetic mutations, identifying pathogens, and studying gene function. The potential for rapid, point-of-care diagnostic tests using CRISPR technology is particularly exciting, offering the possibility of accurate and affordable diagnostics in diverse settings (Broughton et al., 2020).

Microscopy and molecular diagnostics

The study of molecular diagnostics was greatly impacted by the invention of microscopy, which was a turning point in scientific history. The earliest microscopes, pioneered in the late 16th and early 17th centuries by innovators such as Zacharias Janssen and Antonie van Leeuwenhoek, unveiled a hitherto unexplored realm of tiny organisms and cellular structures. Van Leeuwenhoek's careful examination and precise illustrations of bacteria, protozoa, and sperm cells offered the initial insight into the tiny realm of microorganisms, establishing the basis for contemporary microbiology and molecular diagnostics (Dobell, 1932). These initial discoveries were transformative, leading to the eventual development of the germ theory of disease in the 19th century by Louis Pasteur and Robert Koch. This theory, which posited that specific microorganisms cause specific diseases, fundamentally changed medical science. Microscopy enabled the visualization of these microorganisms, providing crucial evidence for the germ theory and revolutionizing the diagnosis and treatment of infectious diseases (Brock, 1999).

During the 19th and early 20th centuries, there were notable improvements in microscopy that enhanced its ability to magnify and provide clearer images. The invention of the compound microscope, which employed many lenses to attain greater magnification, enabled scientists to examine more intricate features within cells. During this time, several staining techniques were introduced, including those pioneered by Paul Ehrlich. These techniques improved the contrast in microscopic pictures and made it easier to study cellular structures and pathogens (Sakula, 1988).

Fluorescence microscopy, which was introduced in the early 20th century, marked a substantial advancement. Fluorescent dyes are employed in this method to mark particular cellular constituents, enabling the observation of structures that would otherwise be imperceptible using a light microscope. Fluorescence microscopy has played a crucial role in the investigation of cell biology, allowing scientists to monitor the actions of specific molecules inside cells. The research has yielded valuable knowledge about cellular processes and disease mechanisms, leading to major progress in molecular diagnostics (Coons et al., 1941).

The invention of the electron microscope in the 1930s by Ernst Ruska and Max Knoll marked another significant advancement in microscopy. Electron microscopes use beams of electrons instead of light to achieve much higher resolutions, allowing scientists to visualize structures at the molecular level. Transmission electron microscopy and scanning electron microscopy have provided detailed images of cellular organelles, viruses, and macromolecular complexes, greatly enhancing our understanding of cellular architecture and function (Ruska, 1986). Electron microscopy has been particularly valuable in the field of virology, enabling the visualization of virus particles and their interactions with host cells.

This has facilitated the study of viral life cycles and the development of antiviral therapies. Additionally, electron microscopy has played a crucial role in structural biology, allowing researchers to determine the three-dimensional structures of proteins and other macromolecules, which is essential for understanding their function and drug development (Knoll & Ruska, 1932). The integration of microscopy with molecular diagnostic techniques has led to groundbreaking advancements in biomedical research and clinical diagnostics. Techniques such as immunohistochemistry and in situ hybridization (ISH) combine microscopy with molecular probes to detect specific proteins or nucleic acid sequences within cells and tissues. These methods have become essential tools for diagnosing cancers, infectious diseases, and genetic disorders (Hsu et al., 1981).

Confocal microscopy, a technique developed in the late 20th century, utilizes laser scanning technology to provide detailed and three-dimensional images of biological samples. This methodology has facilitated comprehensive investigations of the spatial arrangement of cells and tissues, yielding a valuable understanding of intricate biological mechanisms. By combining confocal microscopy with fluorescent probes and molecular markers, researchers have been able to observe and measure molecular interactions in living cells. This has significantly enhanced our knowledge of cellular dynamics and the mechanisms behind diseases (Pawley, 2006).

Fluorescence in situ hybridization (FISH) is a method that utilizes fluorescent imaging and molecular probes to identify certain DNA or RNA sequences within cells. FISH, which was created in the latter part of the 20th century, utilizes probes that are labeled with fluorescent substances. These probes attach to nucleic acid sequences that are complementary, enabling the identification and observation of certain genetic areas. The method is extensively employed in clinical diagnostics to identify chromosomal abnormalities, gene fusions, and other genetic changes linked to different types of malignancies and genetic illnesses (Trask, 1991). FISH has significantly influenced molecular diagnostics by facilitating the accurate identification of genetic defects within cells. This advancement has enhanced the process of diagnosing and categorizing numerous diseases, specifically blood cancers and solid tumors. The capacity to identify precise genetic alterations at the cellular level has also expedited the progress of targeted treatments, thereby enhancing the field of personalized medicine (Gambardella et al., 2020).

The introduction of digital pathology has completely transformed the domains of microscopy and molecular diagnosis. Digital pathology encompasses the process of converting microscopic pictures into digital format and utilizing computational algorithms to evaluate and interpret these images. This methodology allows for the precise measurement and examination of tissue samples, enhancing the precision of diagnoses and facilitating the combination of molecular and structural information (Madabhushi & Lee, 2016). The progress in artificial intelligence (AI) and machine learning is augmenting the capabilities of digital pathology. AI algorithms have the capability to examine extensive collections of digital photos in order to detect patterns and irregularities that could potentially indicate the presence of a disease. This has the potential to enhance the precision and efficiency of diagnosis, mitigate human error, and offer novel insights into disease causes. The incorporation of AI into digital pathology has the potential to revolutionize the domain of molecular diagnostics, presenting novel opportunities for the timely identification and customized management of illnesses (Litjens et al., 2017).

Microarray-based diagnostics

Microarray technology has emerged as an essential tool in molecular diagnostics, facilitating the examination of genetic material in a rapid and efficient manner (**Table 1.3**). This

Table 1.3 Applications and benefits of microarray technology

Application	*Benefits*	*Detailed examples*
Gene expression profiling	Identifies gene activity patterns	Classifying subtypes of breast cancer (e.g., luminal A, luminal B, HER2-enriched, and basal-like) leading to more personalized treatment strategies; identifying prognostic gene signatures in other cancers.
Genotyping and SNP analysis	Identifies genetic predispositions	Pharmacogenomic studies to tailor medications based on individual genetic profiles, such as CYP2C19 polymorphisms affecting clopidogrel metabolism and response in cardiovascular disease patients.
Comparative genomic hybridization	Detects chromosomal imbalances	Identifying chromosomal deletions and duplications in neurodevelopmental disorders like autism and intellectual disabilities; detecting genetic abnormalities in various cancers, aiding in accurate diagnosis and targeted therapy development.
Epigenetic modifications	Studies DNA methylation and histone modifications	Methylation microarrays identifying cancer-specific methylation patterns, serving as biomarkers for early detection, prognosis, and monitoring of disease progression in cancers like colorectal and lung cancer.

method enables the concurrent analysis of several genes, facilitating the investigation of intricate genetic connections and the rapid identification of genetic variants associated with diseases. Microarray technology is mostly used for gene expression profiling (**Figure 1.3**). This technique quantifies the levels of expression of several genes concurrently, offering a momentary overview of gene activity in certain circumstances. Gene expression microarrays have been widely utilized in cancer research to find specific gene signatures linked to various types of cancers. These signatures have the ability to forecast the course of a disease, assist in making treatment choices, and pinpoint possible targets for therapy. Microarray-based gene expression profiling has played a crucial role in categorizing different forms of breast cancer, resulting in the development of more individualized treatment approaches (Perou et al., 2000; van't Veer et al., 2002).

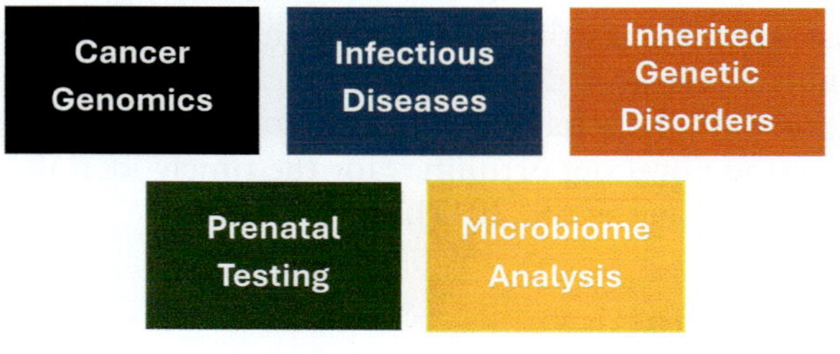

Figure 1.3 Illustration of a microarray method showing probe-target hybridization and signal detection for gene expression analysis.

Microarrays are commonly employed for genotyping and analysis of single nucleotide polymorphisms (SNPs). SNPs are the prevailing kind of genetic variation in individuals, and numerous SNPs are linked to particular diseases or characteristics. Through the analysis of SNPs, scientists are able to detect genetic predispositions to different illnesses and gain insight into the individual variances in how drugs are metabolized and responded. This application is especially helpful in the field of pharmacogenomics since it allows for the customization of pharmaceuticals based on a patient's genetic profile. This customization can improve the effectiveness of the drugs and minimize any negative side effects (Syvänen, 2001; LaFramboise, 2009).

Microarray technology is also used for comparative genomic hybridization (CGH), which is an important application. These microarrays identify chromosomal aberrations, such as deletions and duplications, by comparing the DNA of an individual to a standard genome. This technique is employed to detect genetic anomalies in individuals with congenital diseases and those suffering from cancer. CGH microarrays have played a vital role in identifying chromosomal alterations in neurodevelopmental disorders and different types of malignancies. This has helped in achieving precise diagnoses and the creation of medicines that specifically target these conditions (Pinkel et al., 1998; Cheung et al., 2005).

Microarray technology is used for the investigation of epigenetic alterations, including DNA methylation and histone modifications, which have significant functions in gene regulation. Methylation microarrays can detect methylation patterns throughout the entire genome, offering valuable insights into the regulatory processes that drive changes in gene expression associated with illnesses such as cancer. These patterns can function as biomarkers for the early identification, prediction, and tracking of illness advancement (Bibikova et al., 2006; Laird, 2010).

The progress in microarray technology has resulted in the creation of high-density microarrays, which provide improved resolution and the capacity to study several genetic variations simultaneously. High-density arrays have enhanced the ability to detect infrequent variations and minor genetic alterations that may go unnoticed when using arrays with lower density. The progress has improved the precision and dependability of genetic testing, specifically in intricate disorders that involve several genetic components (Pease et al., 1994; Heller, 2002).

Point-of-care testing and digital health

The growing need for quick and easily accessible medical diagnostics has resulted in the widespread use of point-of-care testing (POCT)—a type of medical diagnostic testing that is conducted at or near the location where the patient is receiving care. It offers quick data, which can be utilized to make well-informed clinical decisions. This method differs from conventional laboratory-based testing, which frequently entails extended processing durations and delayed outcomes (Price, 2001).

POCT devices consist of a range of technologies specifically developed to fulfill certain diagnostic requirements. Common examples are glucometers used to monitor blood glucose levels in the management of diabetes, as well as lateral flow tests like pregnancy tests and fast antigen testing for infectious diseases such as COVID-19. These technologies provide notable benefits in terms of convenience, efficiency, and user-friendliness, allowing healthcare providers to promptly identify and treat patients (Luppa et al., 2011).

The introduction of POCT has had a significant and far-reaching effect on the provision of medical care for patients. POCT devices provide regular monitoring and prompt modifications to treatment protocols, hence enhancing patient outcomes in chronic disease management. Rapid diagnostic testing in acute care settings can accelerate clinical decision-making, decrease hospital stays, and decrease healthcare expenses. Furthermore, POCT has played a crucial role in enhancing healthcare availability in distant and underprivileged regions, where there may be a lack of laboratory facilities (St John & Price, 2014).

Integrating digital health technology with POCT has significantly enhanced the capability and scope of diagnostic testing. Digital health involves a wide array of technologies, such as telemedicine, mHealth applications, wearable devices, and Electronic Health Record (EHRs). These technologies enable the remote monitoring, gathering of data, and communication between patients and healthcare practitioners, hence improving the overall efficiency and efficacy of healthcare delivery (Kvedar, Coye, & Everett, 2014).

Telemedicine enables patients to engage in virtual consultations with healthcare specialists through video conversations, thereby minimizing the necessity for face-to-face appointments and enhancing the accessibility of healthcare services, particularly during the COVID-19 pandemic. Mobile health applications empower users to monitor their health parameters, receive prompts for medication, and access tailored health information. Wearable devices, such as smartwatches and fitness trackers, constantly track important body signals and levels of physical activity. They provide immediate data that can be combined with other health records to effectively manage overall treatment (Bashshur et al., 2016).

The synergy between POCT and digital health technologies is exemplified by the development of connected POCT devices that transmit test results to electronic health records or directly to healthcare providers. This connectivity enhances data accuracy, enables trend analysis, and supports proactive management of chronic conditions. For example, connected glucometers can automatically log blood glucose readings, alerting patients and their caregivers to potential issues and prompting timely interventions (Garg et al., 2010).

POCT and digital health technologies have a profound effect on the management of chronic diseases, which is one of their most notable advantages. Chronic ailments, such as diabetes, hypertension, and cardiovascular illnesses, necessitate continuous surveillance and control to avert complications and uphold a high standard of living. POCT devices empower patients to conduct routine self-monitoring in the comfort of their own homes, thereby decreasing the need for frequent trips to the clinic and facilitating the delivery of individualized healthcare. Digital health technologies facilitate this process by offering platforms for patients to document and exchange their health data with healthcare practitioners. This uninterrupted stream of information allows for more agile and adaptable care regimens. For instance, a patient with diabetes can use a connected glucometer to monitor their blood sugar levels, which are then reviewed by their healthcare provider through an app. The healthcare professional has the ability to modify the patient's medication or dietary advice according to the most recent data, providing improved glycemic control and decreasing the likelihood of problems (Huang et al., 2018).

Recent advances and the impact of COVID-19

The COVID-19 pandemic emphasized the crucial significance of molecular diagnostics in the management of infectious diseases. The pressing demand for prompt and precise diagnostic tests resulted in the widespread acceptance of PCR-based assays and the creation of innovative diagnostic technologies. Methods such as digital PCR and CRISPR-based diagnostics were utilized to identify SARS-CoV-2, the virus accountable for COVID-19. This highlights the flexibility and significance of molecular diagnostics in addressing public health crises (Carter et al., 2020).

The epidemic expedited the progress and adoption of portable and point-of-care diagnostic technologies. These advancements have enhanced the accessibility and efficiency of testing, enabling rapid and effective screening of vast populations. The COVID-19 response has provided valuable insights that are expected to propel breakthroughs in molecular diagnostics, hence improving readiness for future outbreaks (Alkhilaiwi & Khan, 2022).

CONCLUSION

The field of molecular diagnostics has seen important scientific milestones and technological developments, leading to its transformation from a basic understanding of genetics to the application of sophisticated, high-throughput technology. This chapter has emphasized the growth from fundamental breakthroughs in genetics and the structure of DNA to the groundbreaking influence of the HGP, the emergence of NGS, and the extensive utilization of microarray technologies. The combination of these technologies with sophisticated microscopy techniques and the emergence of POCT and digital health breakthroughs have substantially improved the capacity to diagnose, monitor, and treat a broad spectrum of diseases. The future of molecular diagnostics has the potential for enhanced precision, personalization, and efficiency in healthcare by combining traditional and advanced technologies. This advancement in medicine will be specifically tailored to an individual's genetic makeup, ushering in a new era of healthcare.

REFERENCES

Alkhilaiwi, F. A., & Khan, N. U. (2022). Era of molecular diagnostics techniques before and after the COVID-19 pandemic. Current Issues in Molecular Biology, 44(10), 4769–4789. https://doi.org/10.3390/cimb44100325

Avery, O. T., MacLeod, C. M., & McCarty, M. (1944). Studies on the chemical nature of the substance inducing transformation of pneumococcal types: Induction of transformation by a desoxyribonucleic acid fraction isolated from Pneumococcus type III. Journal of Experimental Medicine, 79(2), 137–158. https://doi.org/10.1084/jem.79.2.137

Bamshad, M. J., Ng, S. B., Bigham, A. W., Tabor, H. K., Emond, M. J., Nickerson, D. A., & Shendure, J. (2011). Exome sequencing as a tool for Mendelian disease gene discovery. Nature Reviews Genetics, 12(11), 745–755. https://doi.org/10.1038/nrg3031

Bashshur, R. L., Shannon, G. W., Bashshur, N., & Yellowlees, P. M. (2016). The empirical foundations of telemedicine interventions in primary care. Telemedicine and e-Health, 22(5), 342–375. https://doi.org/10.1089/tmj.2016.0045

Bianchi, D. W., Parker, R. L., Wentworth, J., Madankumar, R., Saffer, C., Das, A. F., ... & Platt, L. D. (2012). DNA sequencing versus standard prenatal aneuploidy screening. New England Journal of Medicine, 367(8), 759–767. https://doi.org/10.1056/NEJMoa1311037

Bibikova, M., Lin, Z., Zhou, L., Chudin, E., Garcia, E. W., Wu, B., ... & Fan, J. B. (2006). High-throughput DNA methylation profiling using universal bead arrays. Genome Research, 16(3), 383–393. https://doi.org/10.1101/gr.4410706

Brock, T. D. (1999). Robert Koch: A Life in Medicine and Bacteriology. ASM Press.

Broughton, J. P., Deng, X., Yu, G., Fasching, C. L., Servellita, V., Singh, J., ... & Chiu, C. Y. (2020). CRISPR-Cas12-based detection of SARS-CoV-2. Nature Biotechnology, 38(7), 870–874. https://doi.org/10.1038/s41587-020-0513-4

Carter, L. J., Garner, L. V., Smoot, J. W., Li, Y., Zhou, Q., Saveson, C. J., ... & Sullivan, D. J. (2020). Assay techniques and test development for COVID-19 diagnosis. ACS Central Science, 6(5), 591–605. https://doi.org/10.1021/acscentsci.0c00501

Cheung, S. W., Shaw, C. A., Yu, W., Li, J., Ou, Z., Patel, A., ... & Lupski, J. R. (2005). Development and validation of a CGH microarray for clinical cytogenetic diagnosis. Genetics in Medicine, 7(6), 422–432. https://doi.org/10.1097/01.GIM.00001710992.63691.32

Cohen, S. N., Chang, A. C. Y., Boyer, H. W., & Helling, R. B. (1973). Construction of biologically functional bacterial plasmids in vitro. Proceedings of the National Academy of Sciences, 70(11), 3240–3244.

Collins F.S., Morgan M., Patrinos A. (2003). The Human Genome Project: lessons from large-scale biology. Science. 300(5617), 286–290. https://doi.org/10.1126/science.1084564

Coons, A. H., Creech, H. J., Jones, R. N., & Berliner, E. (1941). The demonstration of pneumococcal antigen in tissues by the use of fluorescent antibody. Journal of Immunology, 45(3), 159–170.

Dobell, C. (1932). Antony van Leeuwenhoek and his "Little Animals". Harcourt, Brace and Company.

Doudna, J. A., & Charpentier, E. (2014). The new frontier of genome engineering with CRISPR-Cas9. Science, 346, 1258096. https://doi.org/10.1126/science.1258096

Gambardella V, Tarazona N, Cejalvo JM, Lombardi P, Huerta M, Roselló S, Fleitas T, Roda D, Cervantes A. (2020). Personalized Medicine: Recent Progress in Cancer Therapy. Cancers. 12(4), 1009. https://doi.org/10.3390/cancers12041009

Garg, S.K., Zisser, H., Schwartz, S., Bailey, T., Kaplan, R., Ellis, S., Jovanovic, L. (2006). Improvement in Glycemic Control with a Transcutaneous, Real-Time Continuous Glucose Sensor: Randomized Controlled Trial. Diabetes Care. 29, 44–50. https://doi.org/10.2337/diacare.29.01.06.dc05-1686

Goodwin, S., McPherson, J. D., & McCombie, W. R. (2016). Coming of age: Ten years of next-generation sequencing technologies. Nature Reviews Genetics, 17(6), 333–351. https://doi.org/10.1038/nrg.2016.49

Gu, W., Miller, S., & Chiu, C. Y. (2013). Clinical metagenomic next-generation sequencing for pathogen detection. Annual Review of Pathology: Mechanisms of Disease, 14, 319–338. https://doi.org/10.1146/annurev-pathmechdis-012418-012751

Heid CA, Stevens J, Livak KJ, Williams PM. (1996). Real-time quantitative PCR. Genome Research, 6(10), 986–994. https://doi.org/10.1101/gr.6.10.986

Heller, M. J. (2002). DNA microarray technology: Devices, systems, and applications. Annual Review of Biomedical Engineering, 4(1), 129–153. https://doi.org/10.1146/annurev.bioeng.4.020702.153438

Hsu, S. M., Raine, L., & Fanger, H. (1981). Use of avidin-biotin-peroxidase complex (ABC) in immunoperoxidase techniques: A comparison between ABC and unlabeled antibody (PAP) procedures. Journal of Histochemistry & Cytochemistry, 29(4), 577–580. https://doi.org/10.1177/29.4.6166661

Huang, Z., Lum, E., Jimenez, G., Semwal, M., Sloot, P., & Car, J. (2018). Medication management support in diabetes: A systematic assessment of diabetes self-management apps. BMC Medicine, 16(1), 51. https://doi.org/10.1186/s12916-019-1362-1

Knoll, M., & Ruska, E. (1932). The electron microscope. Zeitschrift für Physik, 78, 318–339.

Kvedar, J., Coye, M. J., & Everett, W. (2014). Connected health: A review of technologies and strategies to improve patient care with telemedicine and telehealth. Health Affairs, 33(2), 194–199. https://doi.org/10.1377/hlthaff.2013.0992

LaFramboise, T. (2009). Single nucleotide polymorphism arrays: A decade of biological, computational and technological advances. Nucleic Acids Research, 37(13), 4181–4193. https://doi.org/10.1093/nar/gkp384

Laird, P. W. (2010). Principles and challenges of genome-wide DNA methylation analysis. Nature Reviews Genetics, 11(3), 191–203. https://doi.org/10.1038/nrg2732

Litjens, G., Kooi, T., Bejnordi, B. E., Setio, A. A., Ciompi, F., Ghafoorian, M., ... & van Ginneken, B. (2017). A survey on deep learning in medical image analysis. Medical Image Analysis, 42, 60–88. https://doi.org/10.1016/j.media.2017.07.005

Luppa, P. B., Müller, C., Schlichtiger, A., & Schlebusch, H. (2011). Point-of-care testing (POCT): Current techniques and future perspectives. Trends in Analytical Chemistry, 30(6), 887–898. https://doi.org/10.1016/j.trac.2011.01.019

Madabhushi, A., & Lee, G. (2016). Image analysis and machine learning in digital pathology: Challenges and opportunities. Medical Image Analysis, 33, 170–175. https://doi.org/10.1016/j.media.2016.06.037

Mardis, E. R. (2008). Next-generation DNA sequencing methods. Annual Review of Genomics and Human Genetics, 9, 387–402. https://doi.org/10.1146/annurev.genom.9.081307.164359

Mazzarello, P. (1999). A unifying concept: The history of cell theory. Nature Cell Biology, 1(1), E13–E15. https://doi.org/10.1038/8964.

Metzker, M. (2010). Sequencing technologies—the next generation. Nature Reviews Genetics, 11, 31–46. https://doi.org/10.1038/nrg2626

Meyerson, M., Gabriel, S., & Getz, G. (2010). Advances in understanding cancer genomes through second-generation sequencing. Nature Reviews Genetics, 11(10), 685–696. https://doi.org/10.1038/nrg2841

Mullis K. B., & Faloona F. A. (1987). Specific synthesis of DNA in vitro via a polymerase-catalyzed chain reaction. Methods Enzymol. 155, 335–350. https://doi.org/10.1016/0076-6879(87)55023-6.

Pawley, J. B. (2006). Handbook of Biological Confocal Microscopy. Springer Science & Business Media.
Pease, A. C., Solas, D., Sullivan, E. J., Cronin, M. T., Holmes, C. P., & Fodor, S. P. (1994). Light-generated oligonucleotide arrays for rapid DNA sequence analysis. Proceedings of the National Academy of Sciences of the United States of America, 91(11), 5022–5026. doi: 10.1073/pnas.91.11.5022
Perou, C. M., Sorlie, T., Eisen, M. B., van de Rijn, M., Jeffrey, S. S., Rees, C. A., … & Botstein, D. (2000). Molecular portraits of human breast tumours. Nature, 406(6797), 747–752. https://doi.org/10.1038/35021093
Pinkel, D., Segraves, R., Sudar, D., Clark, S., Poole, I., Kowbel, D., … & Albertson, D. G. (1998). High resolution analysis of DNA copy number variation using comparative genomic hybridization to microarrays. Nature Genetics, 20(2), 207–211. https://doi.org/10.1038/2524
Price, C. P. (2001). Point of care testing. BMJ, 322(7297), 1285–1288. https://doi.org/10.1136/bmj.322.7297.1285
Qin, J., Li, R., Raes, J., Arumugam, M., Burgdorf, K. S., Manichanh, C., … & Wang, J. (2010). A human gut microbial gene catalogue established by metagenomic sequencing. Nature, 464(7285), 59–65.
Ruska, E. (1986). The development of the electron microscope and of electron microscopy. Nobel Lecture, December 8, 1986.
Sakula, A. (1988). Paul Ehrlich: Founder of chemotherapy. Praxis, 77(7), 113–115.
Sanger, F., Nicklen, S., & Coulson, A. R. (1977). DNA sequencing with chain-terminating inhibitors. Proceedings of the National Academy of Sciences of the United States of America, 74(12), 5463–5467. doi: 10.1073/pnas.74.12.5463.
St John, A., & Price, C. P. (2014). Existing and emerging technologies for point-of-care testing. The Clinical Biochemist Reviews, 35(3), 155–167.
Syvänen, A. C. (2001). Accessing genetic variation: Genotyping single nucleotide polymorphisms. Nature Reviews Genetics, 2(12), 930–942. https://doi.org/10.1038/35103535
Trask, B. J. (1991). Fluorescence in situ hybridization: Applications in cytogenetics and gene mapping. Trends in Genetics, 7(5), 149–154.
van't Veer, L. J., Dai, H., van de Vijver, M. J., He, Y. D., Hart, A. A., Mao, M., … & Bernards, R. (2002). Gene expression profiling predicts clinical outcome of breast cancer. Nature, 415(6871), 530–536. https://doi.org/10.1038/415530a
Watson, J., & Crick, F. (1953). Molecular structure of nucleic acids: A structure for deoxyribose nucleic acid. Nature, 171, 737–738. https://doi.org/10.1038/171737a0

Chapter 2

Sample handling in molecular diagnostics

Hanaa Ghabban
Department of Biology, Faculty of Science,
University of Tabuk, Tabuk, Saudi Arabia
Biodiversity Genomics Unit, Faculty of Science,
University of Tabuk, Tabuk, Saudi Arabia

Gourab Paul
Amity School of Biological Sciences,
Amity University Mohali, India

*Irshad Ahmad**
Sir Syed Faculty of Science, Department of Biotechnology,
Mohammad Ali Jauhar University, Rampur, India

INTRODUCTION

The quality and integrity of biological samples are crucial for the efficacy of molecular diagnostics. Ensuring the appropriate gathering, conservation, and conveyance of samples is essential for acquiring dependable and precise diagnostic outcomes. This chapter provides an in-depth exploration of the many types of biological samples, the methods and protocols used for their collection, the recommended practices for sample preservation and transportation, and the quality control measures involved in handling samples.

TYPES OF BIOLOGICAL SAMPLES

The choice of biological sample utilized in molecular diagnostics might have a substantial impact on the results of the tests. Various samples yield diverse amounts of information and are appropriate for distinct diagnostic objectives. Typical biological samples consist of blood, urine, saliva, tissue, swabs, and cerebrospinal fluid (CSF) (**Table 2.1**).

Blood is a highly adaptable and often employed biological specimen in molecular diagnostics. It can be employed to isolate DNA, RNA, and proteins, rendering it indispensable for a diverse array of examinations, such as genetic testing, infectious disease diagnosis, and chronic disease monitoring. The main constituents utilized in these analyses are whole blood, plasma, and serum. Blood samples are commonly favored because they contain a high amount of genetic material and may be collected in a relatively simple manner (Kubista et al., 2006).

Urine samples are highly valuable in identifying urinary tract infections, kidney illnesses, and specific metabolic disorders. They are non-invasive and easily collectible, rendering them ideal for regular screenings and monitoring. Urine can also be utilized for the identification of biomarkers associated with different types of malignancies and hormonal abnormalities. Recent

* Corresponding author.

DOI: 10.1201/9781003489177-2

Table 2.1 Types of biological samples and their diagnostic applications

Sample type	*Applications*	*Benefits*	*Considerations*
Blood	Genetic testing, infectious disease diagnosis, chronic disease monitoring	Rich genetic material, straightforward collection	Requires venipuncture, potential for hemolysis
Urine	Detection of urinary tract infections, kidney diseases, metabolic conditions	Non-invasive, easy to collect	May require preservatives, potential for bacterial growth
Saliva	Genetic testing, hormone level assessment, infectious disease detection	Non-invasive, easy to collect	May require stabilizing agents
Tissue	Cancer diagnosis, organ-specific disease diagnosis	Direct source of affected cells, detailed molecular analysis	Invasive, requires biopsy
Swabs	Respiratory infections, sexually transmitted infections	Easy to collect, rapid processing	Potential for contamination, requires proper aseptic technique
Cerebrospinal fluid	Central nervous system infections, neurological disease diagnosis	Detailed information about CNS conditions	Invasive, requires lumbar puncture

developments have enhanced the ability of molecular diagnostic tests to detect substances in urine, making it a valuable substitute for more intrusive techniques (Bustin et al., 2009).

Saliva is a sort of sample that does not require invasion and is becoming more commonly utilized for genetic testing, hormone level evaluations, and detecting infectious diseases. Collecting this data is straightforward and can yield a plethora of information regarding an individual's health condition. Saliva-based tests are highly effective in detecting viral infections, including HIV and COVID-19, as well as in conducting pharmacogenomic investigations (Mori & Notomi, 2009).

Tissue samples, collected through biopsies, are essential for the diagnosis of malignancies and other disorders that impact certain organs. These samples offer a straightforward and precise source of cells that are damaged, enabling thorough examination and analysis of their histological and molecular characteristics. Tissue biopsies have the ability to detect mutations, changes in gene expression, and other molecular modifications that are crucial for precise diagnosis and planning of treatment (Hindson et al., 2011).

Swabs are frequently employed for the purpose of gathering samples from the nasal cavity, throat, or other mucous membranes. They are crucial for identifying lung infections, sexually transmitted diseases, and other illnesses that impact mucosal tissues. Swab samples are easily collected and may be processed expeditiously, rendering them well-suited for rapid diagnostic tests (Chen et al., 2018).

CSF samples are acquired via lumbar puncture and are utilized for the diagnosis of infections and disorders affecting the central nervous system, such as meningitis and multiple sclerosis. CSF analysis is a helpful diagnostic tool for detecting the presence of infections, inflammatory markers, and other signs of neurological disorders (Gootenberg et al., 2018).

COLLECTION METHODS AND BEST PRACTICES

Ensuring the quality and reliability of diagnostic tests relies heavily on following proper sample-collecting methods and adhering to best practices. The following factors are crucial for efficiently gathering samples:

1. Aseptic method—It is employed to prevent contamination, requiring all sample-collecting processes to be conducted with meticulous care to maintain sterility. This entails employing aseptic equipment, donning gloves, and guaranteeing a hygienic

setting. Contamination can introduce foreign DNA or RNA, resulting in inaccurate positive results or impaired accuracy of the data (Kubista et al., 2006).

2. Accurate identification—It is essential that each sample is distinctly marked with patient details, such as the full name, date of birth, and the precise date and time of collection. Precise labeling guarantees the capacity to track and decreases the likelihood of sample errors, which can have significant consequences for patient diagnosis and treatment (Bustin et al., 2009).
3. Optimal timing—The timing of sample collection can be critical for specific tests. For example, in order to accurately measure hormone levels, it may be necessary to obtain blood samples at specific times of the day to capture the natural variations. Furthermore, the identification of infectious organisms might be influenced by the timing of sample collection, particularly in cases of acute illnesses (Mori & Notomi, 2009).

Whenever feasible, it is advisable to prioritize non-invasive or minimally invasive techniques for sample collection in order to minimize patient discomfort and the potential for consequences. This involves utilizing saliva or urine samples as alternatives to blood or tissue biopsies, when suitable (Hindson et al., 2011).

Preparation of the patient is crucial since it can potentially affect the quality of the sample. These requirements may involve abstaining from food before collecting a blood sample, refraining from taking particular drugs, or adhering to specific instructions for urine collection. Effective communication with the patient regarding these criteria is crucial (Chen et al., 2018).

SAMPLE PRESERVATION AND TRANSPORT

Ensuring the preservation of biological samples from the moment they are collected until they are analyzed is essential for achieving precise diagnostic outcomes. Important factors in the preservation and transportation of samples include:

1. Temperature control is crucial for preserving the quality of different types of samples, as appropriate storage temperatures are necessary to prevent degradation. For instance, RNA is extremely prone to deterioration and should be stored at a temperature of −80°C if it is not going to be processed right away. Typically, blood samples are maintained at a temperature of 4°C in order to maintain the integrity of biological components and avoid the breakdown of red blood cells (hemolysis) (Kubista et al., 2006).
2. Stabilizing agents, such as preservatives, can be added to samples to preserve their integrity during transportation. For example, anticoagulants may be used to treat blood samples, whereas preservatives may be necessary for urine samples to prevent bacterial development (Bustin et al., 2009).
3. Minimizing the time between sample collection and processing is crucial in order to mitigate the risk of degradation. Efficient transportation is crucial for samples that are vulnerable to environmental factors, such as temperature and humidity (Mori & Notomi, 2009).
4. Adequate packaging is crucial to prevent any potential leakage or contamination during transportation. This involves the utilization of biohazard bags, containers that are resistant to leaks, and proper labeling to indicate the contents and any specific handling instructions (Hindson et al., 2011).

QUALITY CONTROL IN SAMPLE HANDLING

Implementing quality control methods is crucial in order to ensure the dependability of diagnostic outcomes. The measures encompass the following:

1. Laboratories must establish and adhere to standard operating procedures (SOPs) for sample collection, preservation, transport, and processing. It is important to document and routinely review these procedures to ensure that they continue to be effective and current (Kubista et al., 2006).
2. Personnel responsible for sample handling must get comprehensive training in the SOPs and possess a thorough understanding of the criticality of preserving sample integrity. Regular training and competency tests are crucial for ensuring strict adherence to best practices (Bustin et al., 2009).
3. Continuous monitoring and comprehensive documentation of the sample handling procedure aid in identifying and correcting any deviations from the SOPs. This includes the monitoring of sample conditions, transportation durations, and other occurrences that could potentially affect the quality of the samples (Mori & Notomi, 2009).
4. Regular validation of collection and handling processes guarantees their effectiveness and the preservation of samples' viability for analysis. Validation studies can detect possible causes of variability and assist in refining protocols to reduce these risks (Hindson et al., 2011).
5. Performing routine internal audits and engaging in external quality evaluation programs can effectively pinpoint areas for enhancement in sample handling procedures. These audits offer an impartial evaluation of the laboratory's compliance with SOPs and optimal methods (Chen et al., 2018).

TROUBLESHOOTING

Effective sample collection is pivotal for accurate molecular diagnostics. However, several common issues can compromise sample integrity and affect diagnostic outcomes. Identifying these issues and implementing troubleshooting strategies is essential.

Common issues in sample collection

Hemolysis in blood samples

Hemolysis is the process of rupturing red blood cells, which leads to the release of hemoglobin into the plasma. This can disrupt a variety of assays, resulting in imprecise outcomes.

Possible causes of hemolysis include incorrect venipuncture technique, excessive force during sample collection, or poor handling and transportation.

Prevention and resolution: In order to reduce the occurrence of hemolysis, it is advisable to employ a delicate method when drawing blood, refrain from using small-gauge needles, and guarantee appropriate handling and storage. To address the presence of hemolysis, it is recommended to collect another sample using appropriate procedures (Kubista et al., 2006).

Contamination

The presence of extraneous DNA or RNA in the sample can occur due to contamination, which can result in inaccurate positive results or impaired integrity of the data.

Possible causes of contamination include the use of non-sterile equipment, insufficiently rigorous aseptic methods, or the unintentional transfer of contaminants between different samples.

Prevention and resolution: It is crucial to always utilize sterile equipment and uphold aseptic methods. Implementing distinct zones for sample collecting and processing can effectively reduce the risk of cross-contamination. If there is a suspicion of contamination, it is advisable to reject the sample and get a new sample using stringent aseptic techniques (Bustin et al., 2009).

Insufficient sample volume

Insufficient sample volume can result in the inability to conduct essential tests or necessitate multiple collections, hence causing discomfort to the patient.

Possible causes of insufficient volume may include incorrect sample collection techniques or inadequate information given to the patient.

Preventive measures and resolution: Guarantee adequate training for staff about sample collection techniques and deliver explicit instructions to patients. If there is not enough volume, it is necessary to collect a new sample to ensure that a suitable volume is achieved (Mori & Notomi, 2009).

Degradation of the genetic material

DNA or RNA can degrade as a result of inadequate storage or extended transportation durations. The degradation of nucleic acids can occur due to factors such as high temperature, exposure to light, or inadequate storage conditions.

Preventive measures and solutions: Ensure that samples are promptly stored at suitable temperatures following collection. Employ stabilizing agents as needed and minimize transportation durations. If there is a suspicion of degradation, it is advisable to obtain a fresh sample and ensure its adequate preservation (Hindson et al., 2011).

Sample mislabeling

Inaccurate labeling of samples might result in incorrect identification and erroneous outcomes.

Causes: Human mistake in the labeling process.

Prevention and resolution: Establish double-check mechanisms in which various people validate the labeling information. Implement barcode technologies to reduce the occurrence of errors that result from human data entry (Chen et al., 2018).

Advanced sample preparation techniques

Microfluidics

The use of microfluidic technologies has revolutionized the field of molecular diagnostics by allowing for the precise manipulation of minuscule amounts of fluids. These devices, commonly known as "labs-on-a-chip," combine various stages of sample processing, including mixing, separation, and reaction, into a single, small device. This integration not only minimizes the loss and contamination of samples but also greatly improves the speed and effectiveness of diagnostic analyses (**Table 2.2**).

Table 2.2 Advanced sample preparation techniques and their benefits

Technique	*Description*	*Benefits*
Microfluidics	Integration of sample processing steps into a compact device	Handles small volumes, reduces sample loss, enables rapid and precise diagnostics
Automated liquid handling	Automated systems for dispensing, transferring, and mixing liquids	High precision and accuracy, reduces human error, enhances throughput and reproducibility
Bead-based methods	Use of functionalized beads to isolate and analyze target molecules	High specificity and efficiency, scalable and automatable, suitable for high-throughput applications

Microfluidics offers a significant benefit in its capacity to manipulate very small quantities of samples and reagents, typically ranging from nanoliters to picoliters. The process of miniaturization decreases the usage of costly substances and enables working with small sample sizes, which is especially advantageous in clinical environments where the availability of samples can be restricted. In addition, the use of small volumes in microfluidic devices allows for quick changes in temperature and effective blending, which are crucial for processes like polymerase chain reaction (PCR) and other amplification techniques (Whitesides, 2006).

Microfluidic systems are extremely adaptable and can be engineered to conduct a diverse array of diagnostic procedures (**Figure 2.1**). As an illustration, they have the capability to combine the processes of extracting, amplifying, and detecting nucleic acids onto a single platform, which allows for smooth and uninterrupted workflows from the first sample to the final result. This feature is particularly advantageous in point-of-care diagnostics, where prompt and dependable outcomes are crucial. Microfluidic devices have been employed in the detection of various infectious diseases, cancer biomarkers, and genetic disorders, demonstrating their broad applicability and potential to revolutionize healthcare (Sackmann et al., 2014).

The combination of microfluidics and sophisticated sensing technologies has augmented its diagnostic capabilities. For instance, microfluidic platforms can be coupled with digital PCR and next-generation sequencing (NGS) technologies, providing highly sensitive and specific detection of genetic mutations, pathogen DNA, and other biomarkers. These

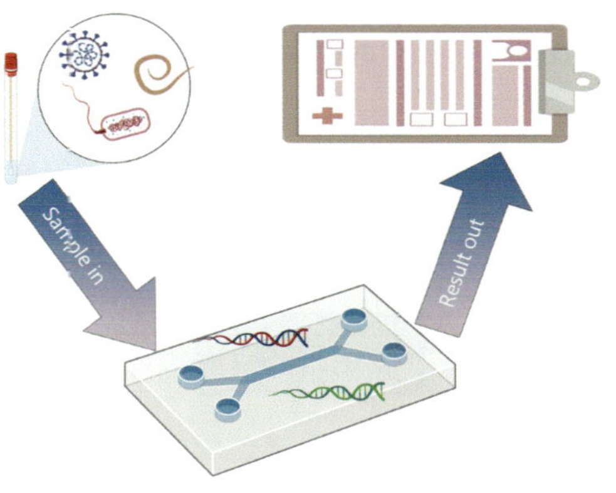

Figure 2.1 A microfluidic system for infectious disease diagnostics, integrating fluid pathways, detection modules, and automated sample processing.

integrated systems allow for the analysis of single cells and rare genetic variants, which are often challenging to detect with conventional methods. The ability to perform single-cell analysis is particularly important in cancer diagnostics, where tumor heterogeneity can significantly impact treatment outcomes (Zhang et al., 2019).

Recent advancements in microfluidic technology have focused on enhancing automation and user-friendliness, making these devices more accessible to non-specialist users. Automated microfluidic systems can perform complex diagnostic workflows with minimal human intervention, reducing the potential for user error and improving reproducibility. Additionally, the development of portable and user-friendly microfluidic devices has expanded the reach of diagnostics to remote and resource-limited settings, where traditional laboratory infrastructure may be unavailable (Whitesides, 2006; Sackmann et al., 2014).

Moreover, microfluidics offers significant potential for high-throughput screening (HTS) applications. By integrating multiple reaction chambers on a single chip, microfluidic systems can perform thousands of parallel assays, making them ideal for large-scale screening of drug candidates, genetic variants, and environmental samples. This high-throughput capability is invaluable for research and development in pharmaceuticals, genomics, and environmental monitoring (Zhang et al., 2019).

Automated liquid handling systems

Automated liquid handling systems have significantly revolutionized the molecular diagnostics sector by substantially improving the efficiency, accuracy, and consistency of sample preparation procedures. These systems are engineered to carry out a range of functions, such as dispensing, transferring, mixing, and diluting liquids, with minimal human involvement (**Figure 2.2**). Automating processes not only accelerates the progression of tasks but also diminishes the likelihood of human mistakes, which is crucial for upholding the integrity and precision of diagnostic outcomes.

An inherent benefit of automated liquid handling systems is their capacity to distribute liquids with exceptional precision and accuracy. These systems employ cutting-edge technologies, including robotic arms, pipetting heads, and sensors, to guarantee precise handling of liquid quantities consistently. Precision is of utmost importance in molecular diagnostics, as the accuracy of liquid volumes can directly influence the sensitivity and specificity of assays (Shih et al., 2012).

Automated pipetting systems have the capability to correctly measure microliter to nanoliter volumes with a high level of reproducibility, guaranteeing precise measurement of each

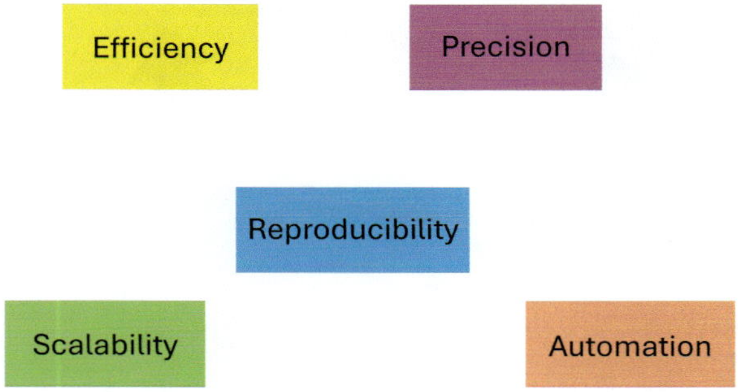

Figure 2.2 Illustration summarizing the core benefits of automated liquid handling systems, including accuracy, throughput, contamination control, and reproducibility.

sample and reagent. Accurate measurement is crucial in applications like quantitative PCR and NGS, as the reliability of the data produced can be influenced by the uniformity of sample and reagent amounts (Yin et al., 2019).

Automated liquid handling systems are specifically engineered to efficiently manage workflows that include a significant volume of samples, making them particularly well-suited for laboratories that handle a huge quantity of samples. These systems can be configured to do repetitive activities on several samples concurrently, so greatly enhancing the efficiency of diagnostic procedures.

In HTS applications, automated liquid handlers have the capability to handle hundreds to thousands of samples per day. This allows for the efficient screening of genetic variations, drug candidates, and other analytes. The capacity to scale is critical for conducting large-scale investigations, such as genome-wide association studies (GWAS) and pharmacogenomic research, where rapid processing of big datasets is necessary (Zhang et al., 2019).

Contemporary automated liquid handling systems have the capability to be combined with other laboratory equipment and software, resulting in the creation of completely automated processes. These integrated systems have the capability to carry out intricate activities, including preparing samples, extracting nucleic acids, amplifying them, and detecting them, all without the need for manual involvement. This degree of automation minimizes the possibility of human mistakes and contamination, guaranteeing that samples are handled consistently and precisely.

For example, the integration of automated liquid handlers with microfluidic devices, heat cyclers, and plate readers can optimize operations for various applications like NGS library preparation, multiplex PCR, and enzyme-linked immunosorbent assay (ELISA). These linked workflows not only increase efficiency but also improve the capacity to reproduce and rely on diagnostic outcomes (Frey, 2015).

Automated liquid handling systems are fitted with sophisticated sensors and software that allow for real-time monitoring and quality control. These characteristics guarantee that any deviations from the set protocols are rapidly identified and rectified, hence preserving the integrity of the samples and the precision of the outcomes.

For instance, liquid handling systems may have functionalities such as liquid level detection, tip alignment verification, and automated calibration procedures. The implementation of these quality control measures guarantees the accurate execution of every stage in the sample preparation process, hence minimizing the possibility of mistakes and enhancing the overall excellence of the diagnostic data (Nir et al., 2020).

Applications in molecular diagnostics

Automated liquid handling systems are extensively utilized in diverse applications within the field of molecular diagnostics, encompassing:

Genomics and transcriptomics: Automated systems are indispensable for the preparation of samples for sequencing, encompassing DNA/RNA extraction, library preparation, and target enrichment.

Proteomics involves the use of automated liquid handlers to perform tasks such as protein extraction, purification, enzymatic digestion, and the setup of high-throughput immunoassays.

Clinical diagnostics benefit from the use of automated technologies, which optimize processes in laboratories by performing tasks such as PCR setup, ELISA, and sample aliquoting. This implementation enhances efficiency and decreases the time required for completion (Yin et al., 2019).

Bead-based methods

Bead-based methods are versatile and effective procedures commonly employed in molecular diagnostics to isolate, purify, and analyze biological molecules like DNA, RNA, proteins, and cells. These approaches utilize the surface features of beads, which may be modified with specific ligands to trap target molecules with a high degree of specificity and efficiency. these techniques utilize magnetic or non-magnetic beads that are coated with ligands, such as antibodies, oligonucleotides, or other binding molecules, which selectively attach to the desired analytes. After the target molecules are trapped, the beads can be effortlessly isolated from the mixture by employing magnetic fields (for magnetic beads) or centrifugation (for non-magnetic beads). This segregation enables the focusing and refinement of the desired molecules, rendering them appropriate for subsequent investigation (Qiu et al., 2024).

Applications in molecular diagnostics

Nucleic acid purification is a frequently used use of bead-based techniques. Magnetic beads, which have been coated with silica or other ligands that attach to nucleic acids, are employed to specifically capture DNA or RNA from intricate biological samples. This procedure exhibits a high level of effectiveness and has the capability to be mechanized, rendering it well-suited for tasks that require a large volume of work to be completed in a short amount of time, such as NGS and PCR (Hawkins et al., 1994).

Immunoassays, such as ELISA, employ beads that are coated with antibodies to selectively capture certain antigens or antibodies from samples. These assays exhibit a high level of sensitivity and specificity, rendering them valuable for the detection of proteins, pathogens, and biomarkers present in low quantities in clinical samples (Anderson & Hunter, 2006).

Bead-based techniques are utilized for the separation and examination of particular cell populations in the process of cell sorting and analysis. Magnetic-activated cell sorting is a technique that uses magnetic beads coated with antibodies to selectively collect and separate particular cell populations from mixtures containing different types of cells. The utilization of this method is prevalent in the fields of immunology, cancer research, and stem cell research (Miltenyi et al., 1990).

Protein purification: Bead-based methods are used for the purification of proteins through affinity chromatography. Beads coated with specific ligands, such as antibodies or metal ions, can capture target proteins from complex mixtures, enabling their purification and concentration for further analysis (Dutta & Bose, 2022).

Advantages

High specificity and efficiency: Bead-based methods offer high specificity and efficiency due to the customizable surface chemistry of the beads. This allows for the selective capture of target molecules while minimizing the presence of contaminants.

Scalability and automation: These methods are highly scalable and can be easily automated, making them suitable for high-throughput applications. Automated bead-based platforms are widely used in clinical laboratories and research settings for large-scale sample processing.

Versatility: These methods can be adapted for a wide range of applications, including nucleic acid purification, protein isolation, immunoassays, and cell sorting. This versatility makes them valuable tools in molecular diagnostics and biomedical research.

Rapid processing: The ability to quickly separate and purify target molecules using magnetic fields or centrifugation enables rapid processing of samples, which is critical for timely diagnostic results.

Recent innovations

Integration with microfluidics: Recent progress has been centered around combining bead-based techniques with microfluidic devices to improve the effectiveness and automation of sample handling. Microfluidic platforms have the ability to handle small amounts of fluids and beads, allowing for precise control over the collection and release of target molecules. This integration enables the creation of small, portable diagnostic devices for testing at the point of service (Pamme, 2006).

Advanced surface modifications: Advancements in the field of bead surface chemistry have resulted in the creation of beads that have improved the ability to attach to target molecules and decreased the tendency to bind non-specifically. The enhancements mentioned in the study by Hawkins et al. (1994) improve the effectiveness of capturing the desired target and decrease unwanted background noise. This leads to an improvement in the accuracy and precision of bead-based assays.

Multiplexing capabilities: Bead-based techniques have been modified to enable multiplexing, which permits the concurrent identification of many targets in a single test. This feature is especially valuable in diagnostic applications that need simultaneous analysis of many biomarkers. Multiplex bead-based assays, such as Luminex xMAP technology, employ beads with unique optical characteristics to detect many analytes concurrently, hence enhancing the speed and effectiveness of diagnostic procedures (Dunbar, 2012).

CONCLUSION

The quality and integrity of biological samples are of utmost importance in molecular diagnostics to ensure accurate diagnostic outcomes. This chapter discusses the critical aspects of proper sample handling, from collection to analysis. It emphasizes the importance of using aseptic techniques, accurately identifying samples, adhering to precise timing, and minimizing invasiveness to preserve the integrity of the samples. Efficient preservation and transportation techniques are crucial for preserving the viability of samples. Implementing quality control measures, such as defined operating procedures, continuous monitoring, and regular training, is crucial to guarantee accurate diagnostic outcomes. Laboratories can greatly boost precision, throughput, and consistency in diagnostic processes, resulting in improved accuracy and better patient care, by integrating sophisticated technology such as microfluidics and automated liquid handling systems.

REFERENCES

Anderson, L., & Hunter, C. L. (2006). Quantitative mass spectrometric multiple reaction monitoring assays for major plasma proteins. Molecular & Cellular Proteomics, 5(4), 573–588. https://doi.org/10.1074/mcp.M500331-MCP200

Bustin, S. A., Benes, V., Garson, J. A., Hellemans, J., Huggett, J., Kubista, M., ... Wittwer, C. T. (2009). The MIQE guidelines: minimum information for publication of quantitative real-time PCR experiments. Clinical Chemistry, 55(4), 611–622.

Chen, J. S., Ma, E., Harrington, L. B., Da Costa, M., Tian, X., Palefsky, J. M., & Doudna, J. A. (2018). CRISPR-Cas12a target binding unleashes indiscriminate single-stranded DNase activity. Science, 360(6387), 436–439.

Dunbar, S. A. (2012). Bead-Based Suspension Arrays for the Detection and Identification of Respiratory Viruses. In: Tang, Y. W., Stratton, C. (eds.), Advanced Techniques in Diagnostic Microbiology. Springer, 813–833. https://doi.org/10.1007/978-1-4614-3970-7_42

Dutta, S., Bose, K. (2022). Protein Purification by Affinity Chromatography. In: Bose, K. (ed.), Textbook on Cloning, Expression and Purification of Recombinant Proteins. Springer, 141–171. https://doi.org/10.1007/978-981-16-4987-5_6

Frey, O. (2015). Automating genomic workflows with liquid handling systems. Genetic Engineering & Biotechnology News, 35(14), 6–7.

Gootenberg, J. S., Abudayyeh, O. O., Kellner, M. J., Joung, J., Collins, J. J., & Zhang, F. (2018). Multiplexed and portable nucleic acid detection platform with Cas13, Cas12a, and Csm6. Science, 360(6387), 439–444.

Hawkins, T. L., O'Connor-Morin, T., Roy, A., & Santillan, C. (1994). DNA purification and isolation using a solid-phase. Nucleic Acids Research, 22(21), 4543–544. https://doi.org/10.1093/nar/22.21.4543

Hindson, B. J., Ness, K. D., Masquelier, D. A., Belgrader, P., Heredia, N. J., Makarewicz, A. J., ... Colston, B. W. (2011). High-throughput droplet digital PCR system for absolute quantitation of DNA copy number. Analytical Chemistry, 83(22), 8604–8610.

Kubista, M., Andrade, J. M., Bengtsson, M., Forootan, A., Jonák, J., Lind, K., ... Zoric, N. (2006). The real-time polymerase chain reaction. Molecular Aspects of Medicine, 27(2-3), 95–125.

Miltenyi, S., Müller, W., Weichel, W., & Radbruch, A. (1990). High gradient magnetic cell separation with MACS. Cytometry, 11(2), 231–238. https://doi.org/10.1002/cyto.990110203

Mori, Y., & Notomi, T. (2009). Loop-mediated isothermal amplification (LAMP): a rapid, accurate, and cost-effective diagnostic method for infectious diseases. Journal of Infection and Chemotherapy, 15(2), 62–69.

Nir, D., Hasse, E., Zeng, H., & Schlenke, P. (2020). Advanced technologies in liquid handling. SLAS Technology, 25(3), 193–199.

Pamme, N. (2006) Magnetism and microfluidics. Lab on a Chip, 6, 24–38. https://doi.org/10.1039/B513005K

Qiu, X., Zhao, H., Xu, P., & Li, J. (2024). Research on a calculation model of ankle-joint-torque-based sEMG. Sensors, 24(9), 2906. https://doi.org/10.3390/s24092906

Sackmann, E. K., Fulton, A. L., & Beebe, D. J. (2014). The present and future role of microfluidics in biomedical research. Nature, 507(7491), 181–189.

Shih, S. C., Goyal, G., Kim, P. W., Koutsoubelis, N., Keasling, J. D., Adams, P. D., & Hillson, N. J. (2012). A versatile microfluidic device for automating synthetic biology. ACS Synthetic Biology, 1(6), 228–236.

Whitesides, G. M. (2006). The origins and the future of microfluidics. Nature, 442(7101), 368–373.

Yin, H., Kwan, P. H., & Li, X. (2019). Robotic pipetting and liquid handling platforms for lab automation. BioTechniques, 66(4), 190–194.

Zhang, H., Liu, D., & Zhang, X. (2019a). Advances in Liquid Handling for High-Throughput Screening. Journal of Laboratory Automation, 24(1), 5–14.

Zhang, Y., Nguyen, N. T., & Tan, S. H. (2019b). Fundamentals and applications of inertial microfluidics: a review. Lab on a Chip, 19(1), 116–135.

Chapter 3

Polymerase chain reaction and its variants in molecular diagnostics

Anam Nayab

Department of Biosciences and Technology, Emerson University, Multan, Pakistan

Ahmed Waqas

Department of Zoology, Emerson University, Multan, Pakistan

Musharraf Jelani

Rare Diseases Genetics and Genomics, Centre for Omic Sciences, Islamia College, Peshawar, Pakistan

*Muhammad Umair**

Medical Genomics Research Department, King Abdullah International Medical Research Center (KAIMRC), King Saud Bin Abdulaziz University for Health Sciences, Ministry of National Guard Health Affairs (MNGH), Riyadh, Saudi Arabia

INTRODUCTION

The polymerase chain reaction (PCR) has revolutionized molecular diagnostics since its invention by Kary Mullis in 1983. This technique allows for the exponential amplification of specific DNA sequences, making it possible to detect even minute quantities of nucleic acids in a sample. Over the years, numerous variants of PCR have been developed to enhance its sensitivity, specificity, and range of applications in molecular diagnostics.

Principle of PCR

PCR is a molecular biology technique used to amplify specific DNA sequences, making it possible to generate millions of copies of a particular segment of DNA from a small initial amount. The principle of PCR is based on the natural process of DNA replication, but it is carried out in vitro (in a laboratory setting) using a thermal cycler to control the reaction's temperature. The process involves three main steps: denaturation, annealing, and extension, which are repeated for 20–40 cycles to achieve exponential amplification of the target DNA sequence (**Figure 3.1**).

Denaturation

The first step in PCR is denaturation, where the double-stranded DNA template is heated to 94–98°C. This high temperature breaks the hydrogen bonds between the complementary bases of the DNA strands, resulting in the separation of the double-stranded DNA into two single strands. Denaturation is essential because it provides single-stranded DNA templates for the primers to bind to during the next step (Innis & Gelfand, 1990).

* **Corresponding author.**

DOI: 10.1201/9781003489177-3

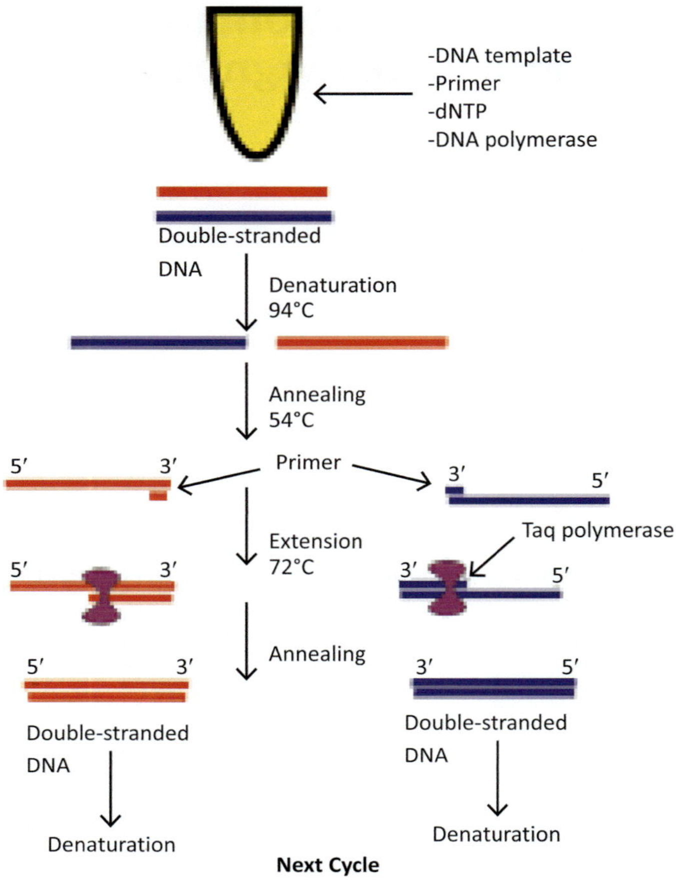

Figure 3.1 Diagram illustrating the Polymerase Chain Reaction (PCR) process showing denaturation, primer annealing, and extension cycles with Taq polymerase.

Annealing

Following denaturation, the temperature is lowered to 50–65°C during the annealing step. At this temperature, short single-stranded DNA sequences known as primers bind, or anneal, to their complementary sequences on the single-stranded DNA template. Primers are designed to flank the target DNA region to be amplified. The specific annealing temperature depends on the melting temperature (Tm) of the primers, which is influenced by their length and nucleotide composition (Erlich, 1989).

Extension

The final step is extension (or elongation), which occurs at 72°C. During this step, the DNA polymerase enzyme synthesizes a new DNA strand by adding nucleotides to the primer-bound template. The most commonly used DNA polymerase in PCR is Taq polymerase, derived from the thermophilic bacterium *Thermus aquaticus*, which is stable and active at high temperatures. Taq polymerase extends the DNA strand from the 3′ end of each primer, incorporating the four deoxynucleoside triphosphates (dNTPs: dATP, dTTP, dCTP, and dGTP) into the growing DNA strand (Saiki et al., 1988).

Thermal cycling

PCR involves repeated cycles of denaturation, annealing, and extension to achieve exponential amplification of the target DNA. Each cycle doubles the number of DNA molecules, leading to a geometric progression in the amount of target DNA. Typically, 20–40 cycles are sufficient to produce millions of copies of the target sequence from a single or few initial copies.

Components of PCR

The PCR reaction mixture contains several critical components:

Template DNA: The DNA sample that contains the target sequence to be amplified.
Primers: Short single-stranded DNA oligonucleotides that are complementary to the target DNA regions flanking the sequence to be amplified.
DNA polymerase: An enzyme that synthesizes new DNA strands by adding nucleotides to the primers. Taq polymerase is commonly used due to its heat stability.
dNTPs: The building blocks (adenine, guanine, cytosine, and thymine) that the DNA polymerase uses to synthesize the new DNA strand.
Buffer solution: Provides the necessary ionic environment and maintains the optimal pH for the DNA polymerase activity.
Magnesium ions: Serve as a cofactor for the DNA polymerase, enhancing its enzymatic activity (Mullis et al., 1986).

Variants of PCR

Real-time PCR (qPCR)

Real-time PCR (qPCR), also known as quantitative PCR, has become an essential tool in molecular diagnostics. This advanced variant of the conventional PCR technique allows for the quantification of DNA and RNA in real time, providing valuable quantitative data for a wide range of diagnostic applications. The ability to monitor the amplification of target nucleic acids as it occurs has revolutionized how scientists and clinicians detect, quantify, and study genetic material.

The principle of qPCR is based on the same basic steps as conventional PCR: denaturation, annealing, and extension. However, qPCR incorporates fluorescent dyes or probes that emit fluorescence in proportion to the amount of PCR product generated during each cycle. This fluorescence is measured in real time, allowing for the quantification of the target nucleic acids throughout the amplification process (**Figure 3.2**).

In qPCR, the fluorescence signal increases as the amount of PCR product increases. The point at which the fluorescence signal exceeds a certain threshold (cycle threshold, or Ct value) is inversely proportional to the initial quantity of the target nucleic acid. This allows for the quantification of the target DNA or RNA in the sample.

Fluorescent chemistries

qPCR utilizes fluorescence-based detection to monitor DNA amplification in real time, offering quantitative insights into the amount of target nucleic acids present in a sample. Two primary types of fluorescent chemistries are employed in qPCR: DNA-binding dyes and sequence-specific probes. Each type has distinct properties, advantages, and applications in

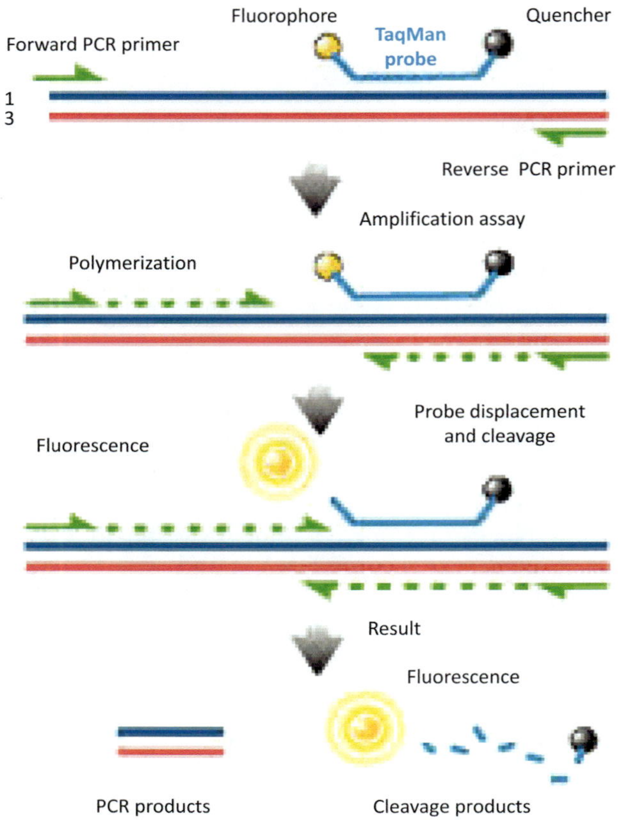

Figure 3.2 Visual of a real-time PCR assay demonstrating probe binding, polymerization, displacement, cleavage, and fluorescence detection.

molecular diagnostics, making them suitable for different purposes based on the specificity and sensitivity required.

DNA-binding dyes

DNA-binding dyes, such as SYBR Green, are intercalating agents that fluoresce upon binding to double-stranded DNA (dsDNA). These dyes offer a straightforward and cost-effective method for detecting DNA amplification. DNA-binding dyes work by intercalating into the minor groove of dsDNA. When these dyes bind to dsDNA, their fluorescence intensity increases dramatically, providing a signal that can be detected and measured during the qPCR process (**Figure 3.3**). The increase in fluorescence is directly proportional to the amount of dsDNA produced during the amplification cycles (Kubista et al., 2006).

ADVANTAGES

The use of DNA-binding dyes simplifies the qPCR setup, as the only additional reagent required is the dye itself. This reduces the complexity of assay design and preparation, making it accessible for various applications.

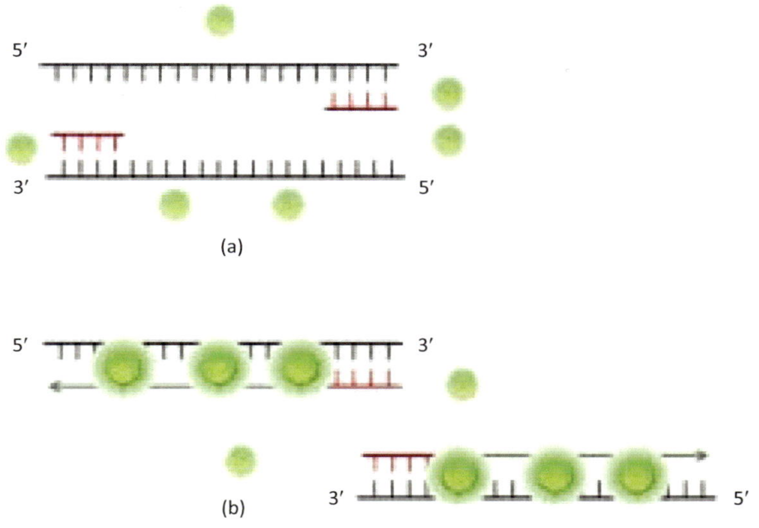

Figure 3.3 Illustration of a DNA sequencing reaction highlighting fluorescent nucleotide incorporation and detection during primer extension.

Cost-effectiveness

DNA-binding dyes are generally less expensive than sequence-specific probes, making them an attractive option for high-throughput applications and routine diagnostics where budget constraints are a consideration.

Broad applicability

DNA-binding dyes can be used for any target DNA sequence without the need for designing specific probes. This versatility makes them suitable for a wide range of qPCR applications, including gene expression analysis and pathogen detection (Nolan et al., 2006).

LIMITATIONS

Non-specific binding

A significant limitation of DNA-binding dyes is their non-specific binding to all dsDNA, including non-specific PCR products and primer dimers. This can lead to false-positive signals and affect the accuracy of quantification. Melting curve analysis is often required to differentiate specific products from non-specific products and primer dimers (Wong & Medrano, 2005). It involves gradually increasing the temperature and monitoring the fluorescence to determine the melting temperature (Tm) of the PCR products. Specific products typically have distinct melting temperatures compared to non-specific products, allowing for their identification (Wong & Medrano, 2005).

Sequence-specific probes

Sequence-specific probes, including TaqMan probes, Molecular Beacons, and Scorpions, are designed to hybridize specifically to the target DNA sequence, providing highly specific and sensitive detection. Sequence-specific probes consist of a fluorescent reporter dye and a quencher. The reporter emits fluorescence when excited, but this fluorescence is quenched by the quencher when the probe is intact. During PCR amplification, the DNA polymerase cleaves the probe bound to the target sequence, separating the reporter from the quencher and resulting in increased fluorescence. This fluorescence signal is proportional to the amount of target DNA, allowing for precise quantification (Heid et al., 1996).

Types of sequence-specific probes

TaqMan PROBES

These hydrolysis probes are cleaved by the 5′ to 3′ exonuclease activity of Taq DNA polymerase during the extension phase (**Figure 3.4**). The cleavage separates the reporter from the quencher, resulting in a fluorescence signal that is directly proportional to the amount of target DNA (Holland et al., 1991).

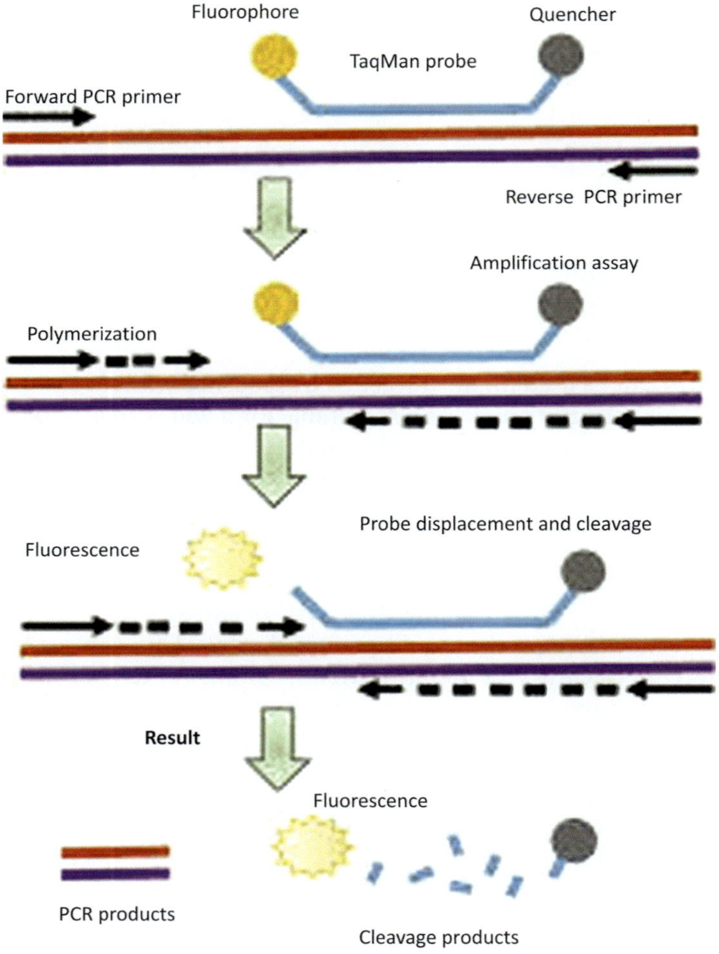

Figure 3.4 Diagram of the TaqMan probe-based PCR process showing amplification, probe cleavage, and resulting fluorescence signal.

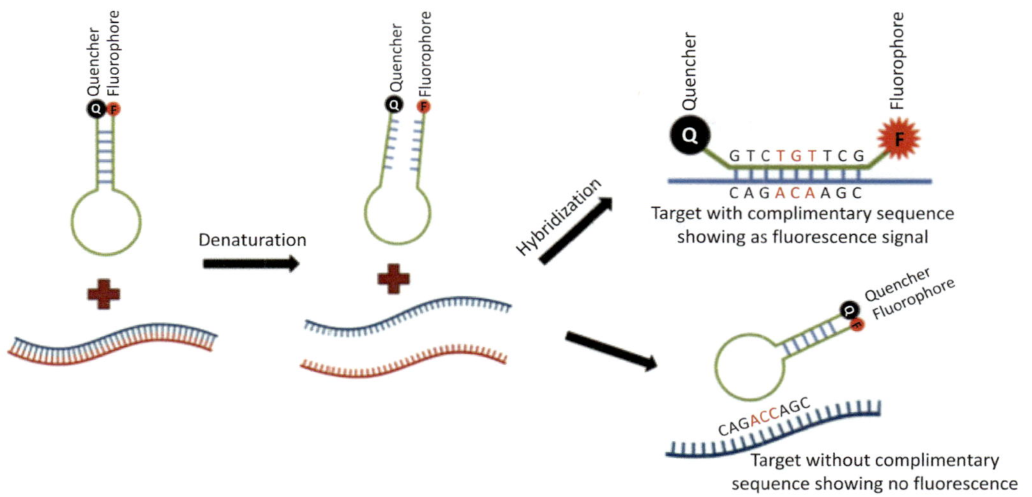

Figure 3.5 Schematic of a hybridization probe system showing denaturation, hybridization, and fluorescence emission upon specific target binding.

MOLECULAR BEACONS

These are hairpin-shaped probes with a reporter at one end and a quencher at the other. In the absence of the target, the probe forms a hairpin loop that brings the quencher close to the reporter, quenching the fluorescence (**Figure 3.5**). Upon hybridization to the target sequence, the hairpin structure opens, separating the quencher from the reporter and emitting fluorescence (Tyagi & Kramer, 1996).

SCORPIONS

Scorpions are probes that contain both the primer and the probe sequences within a single molecule. After the primer extension, the probe portion hybridizes to the target sequence within the same strand, causing the reporter to be separated from the quencher and emitting fluorescence (Whitcombe et al., 1999).

Advantages

HIGH SPECIFICITY

Sequence-specific probes provide high specificity as they only generate a fluorescence signal when hybridized to their target sequence. This reduces the risk of detecting non-specific products and primer dimers.

MULTIPLEXING CAPABILITY

Different probes can be labeled with distinct fluorescent dyes, allowing for the simultaneous detection of multiple targets in a single reaction (multiplexing). This is particularly useful for detecting multiple pathogens or genetic mutations in a single assay (Kubista et al., 2006).

QUANTITATIVE ACCURACY

Sequence-specific probes provide more accurate quantification compared to DNA-binding dyes due to their high specificity and reduced background fluorescence. This is essential for applications requiring precise quantification, such as clinical diagnostics and research studies (Heid et al., 1996).

Applications in molecular diagnostics

Infectious disease detection

qPCR has revolutionized the field of infectious disease diagnostics, offering unparalleled sensitivity and specificity for detecting and quantifying pathogens in clinical samples. This technology is extensively used to diagnose viral infections such as human immunodeficiency virus (HIV), hepatitis B and C, and COVID-19, among others. The ability to quantify viral load is crucial for monitoring disease progression, assessing treatment efficacy, and guiding clinical decision-making.

VIRAL INFECTIONS

HIV: qPCR is used for the diagnosis and management of HIV. It is used to detect and quantify HIV RNA in blood samples, providing crucial information about viral load. This data is vital for initiating antiretroviral therapy, monitoring treatment response, and detecting viral rebound or resistance (Bustin et al., 2009). The ability to measure viral load helps in tailoring treatment regimens to individual patients, improving clinical outcomes.

Hepatitis B and C: qPCR is widely used to detect and quantify hepatitis B virus (HBV) DNA and hepatitis C virus (HCV) RNA. These viruses cause chronic infections that can lead to liver cirrhosis and hepatocellular carcinoma. Monitoring viral load in patients with HBV or HCV infection is crucial for determining the stage of the disease, guiding treatment decisions, and assessing the effectiveness of antiviral therapies (Lok et al., 2001). qPCR enables early detection of viral relapse and helps in adjusting treatment plans accordingly.

COVID-19 (SARS-CoV-2): The COVID-19 pandemic highlighted the importance of qPCR in infectious disease diagnostics. qPCR assays targeting specific genes of the SARS-CoV-2 virus are the gold standard for diagnosing COVID-19. These assays provide rapid and accurate detection of viral RNA in respiratory samples, facilitating timely isolation and treatment of infected individuals. Additionally, qPCR is used to monitor viral load in patients, aiding in the management of severe cases and the assessment of viral clearance (Corman et al., 2020).

BACTERIAL AND FUNGAL INFECTIONS

Tuberculosis (TB): qPCR is employed to detect and quantify *Mycobacterium tuberculosis*, the causative agent of TB. Rapid and accurate diagnosis of TB is essential for initiating appropriate treatment and preventing the spread of the disease. qPCR assays can detect TB bacteria in sputum samples with high sensitivity and specificity, even in cases with low bacterial load (Siddiqi et al., 2012).

Staphylococcus aureus and MRSA: qPCR is used to detect and differentiate between methicillin-sensitive *S. aureus* (MSSA) and methicillin-resistant *S. aureus* (MRSA). This distinction is critical for infection control and selecting appropriate antibiotic therapy. qPCR assays targeting the *mecA* gene, responsible for methicillin resistance, provide rapid results that are crucial for managing hospital-acquired infections (Becker et al., 2004).

Candida infections: qPCR is also applied in the detection of fungal pathogens such as Candida species. Rapid identification and quantification of Candida DNA in clinical samples help in diagnosing invasive candidiasis, a serious infection in immunocompromised patients. qPCR enables early initiation of antifungal therapy, improving patient outcomes (White et al., 2011).

Genetic testing and mutation detection

qPCR is an invaluable tool in the realm of genetic testing and mutation detection. Its high sensitivity and specificity make it particularly suitable for identifying genetic mutations, single nucleotide polymorphisms (SNPs), and chromosomal aberrations. This technology is also extensively utilized in prenatal testing to detect genetic disorders such as Down syndrome. The ability to detect low-frequency mutations in heterogeneous samples further enhances its utility in clinical and research settings.

IDENTIFYING GENETIC MUTATIONS

Genetic mutations are changes in the DNA sequence that can lead to various diseases, including cancer, hereditary disorders, and other genetic conditions. qPCR is widely used to identify these mutations due to its ability to amplify specific DNA sequences and quantify the amount of target nucleic acids present in a sample. By using sequence-specific probes or primers designed to target known mutation sites, qPCR can detect even a single nucleotide change with high precision (Kubista et al., 2006).

One common application of qPCR in mutation detection is in the identification of mutations in oncogenes and tumor suppressor genes. For example, mutations in the *BRCA1* and *BRCA2* genes are associated with an increased risk of breast and ovarian cancers. qPCR assays can be designed to specifically detect these mutations, providing valuable information for risk assessment and management (Bustin et al., 2009).

SINGLE NUCLEOTIDE POLYMORPHISMS

SNPs are the most common type of genetic variation among people. They occur when a single nucleotide in the genome is altered. While most SNPs do not cause disease, some can influence how individuals respond to drugs, susceptibility to environmental factors, and the risk of developing certain diseases. qPCR is particularly effective in detecting SNPs due to its high sensitivity and specificity.

Copy number variation (CNV) analysis: qPCR is commonly used to analyze CNVs, which are variations in the number of copies of a particular gene or genomic region. CNVs can have significant impacts on gene expression and phenotypic traits. By comparing the quantity of the target DNA in a sample to a reference sample, qPCR can determine whether there is a deletion, duplication, or normal copy number. This technique is essential for diagnosing conditions such as DiGeorge syndrome and other genomic disorders (Zhang et al., 2009).

FUSION GENE DETECTION

In cancers like chronic myeloid leukemia (CML), fusion genes such as BCR-ABL1 are formed due to chromosomal translocations. qPCR can be used to detect and quantify these fusion genes, providing critical information for diagnosis, prognosis, and monitoring of treatment response (Cross et al., 1993).

PRENATAL TESTING

Prenatal testing involves screening for genetic disorders in a fetus. qPCR plays a crucial role in non-invasive prenatal testing by analyzing cell-free fetal DNA (cffDNA) present in the maternal blood. This method is used to detect chromosomal aneuploidies such as Down syndrome (trisomy 21), Edwards syndrome (trisomy 18), and Patau syndrome (trisomy 13).

GENE EXPRESSION ANALYSIS

qPCR is a powerful technique widely used in gene expression analysis. This application is essential for studying gene regulation, understanding disease mechanisms, and developing new therapies. qPCR quantifies gene expression levels by measuring the amount of specific RNA transcripts present in a sample. This process involves converting RNA into complementary DNA (cDNA) using reverse transcription, followed by amplification and quantification of the cDNA.

GENE REGULATION STUDIES

qPCR is extensively used to study gene regulation by quantifying the expression levels of genes under different conditions. Researchers can compare gene expression profiles between treated and untreated samples, healthy and diseased tissues, or different developmental stages. This information helps identify regulatory genes and pathways involved in various biological processes (Bustin et al., 2009).

UNDERSTANDING DISEASE MECHANISMS

Alterations in gene expression are often associated with diseases such as cancer, cardiovascular disorders, and neurological conditions. qPCR enables researchers to study these changes by quantifying the expression of disease-related genes. For example, overexpression of oncogenes or underexpression of tumor suppressor genes can be detected in cancerous tissues, providing insights into the molecular mechanisms driving tumor development and progression (Kubista et al., 2006).

BIOMARKER DISCOVERY AND VALIDATION

Gene expression profiling using qPCR is crucial for discovering and validating biomarkers for disease diagnosis, prognosis, and treatment response. Biomarkers are genes whose expression levels correlate with specific disease states or treatment outcomes. For instance, qPCR can identify biomarkers for early cancer detection or predict the response to targeted therapies, enabling personalized medicine (Wong & Medrano, 2005).

THERAPEUTIC DEVELOPMENT

qPCR plays a vital role in developing new therapies by assessing the effects of potential drugs on gene expression. By comparing the expression profiles of treated and untreated cells, researchers can determine the molecular impact of drug candidates. This information is essential for optimizing drug efficacy and safety (Nolan et al., 2006).

ONCOLOGY

qPCR is a pivotal tool in cancer diagnostics, facilitating the detection and quantification of tumor DNA in blood or tissue samples. This technology helps identify specific genetic mutations and chromosomal translocations associated with various cancers. By measuring circulating tumor DNA (ctDNA), qPCR plays a critical role in monitoring minimal residual disease (MRD) and detecting early relapse, thereby improving patient outcomes and enabling personalized treatment strategies.

DETECTION AND QUANTIFICATION OF TUMOR DNA

Tumor DNA: In cancer diagnostics, qPCR is used to detect and quantify DNA from tumor cells, which can be present in blood (as ctDNA) or in tissue biopsies. This process involves designing primers and probes specific to the genetic alterations characteristic of the tumor. The high sensitivity and specificity of qPCR allow for the detection of even small amounts of tumor DNA, making it possible to identify genetic mutations and translocations associated with cancer (Bustin et al., 2009).

CIRCULATING TUMOR DNA

ctDNA refers to fragments of DNA released from tumor cells into the bloodstream. qPCR enables the detection and quantification of ctDNA, providing a non-invasive method for monitoring tumor dynamics. This is particularly useful for patients where traditional biopsy methods are not feasible. By analyzing ctDNA, clinicians can gain insights into the genetic landscape of the tumor and track changes over time (Vogelstein et al., 2013).

Identification of specific mutations and translocations

GENETIC MUTATIONS

Many cancers are driven by specific genetic mutations, such as those in the Kirsten Rst Sarcoma Viral Oncogene Homolog (KRAS), Epidermal Growth Factor Receptor (EGFR), and B-Raf proto-oncogene (BRAF) genes. qPCR is used to detect these mutations with high precision. For example, the presence of KRAS mutations can influence the choice of targeted therapies in colorectal cancer, while EGFR mutations are critical for selecting appropriate treatments in non-small cell lung cancer (Pao & Ladanyi, 2007).

CHROMOSOMAL TRANSLOCATIONS

Certain cancers, such as leukemia and lymphoma, are characterized by specific chromosomal translocations that result in the formation of fusion genes. One well-known example is the BCR-ABL1 fusion gene in CML. qPCR can detect these translocations by amplifying the fusion gene sequences, providing a molecular diagnosis, and guiding treatment decisions (Cross et al., 1993).

MONITORING MINIMAL RESIDUAL DISEASE

MRD refers to the small number of cancer cells that may remain in a patient's body after treatment and can lead to relapse. qPCR is highly sensitive and can detect these residual cancer cells by identifying specific genetic markers associated with the tumor. Monitoring MRD using qPCR allows for early intervention before clinical relapse occurs, improving long-term outcomes for patients (Branford et al., 2014).

FORENSIC SCIENCE

qPCR is an essential tool in forensic science, playing a critical role in analyzing DNA samples collected from crime scenes. qPCR provides a reliable method for quantifying DNA, determining levels of DNA degradation, and assessing sample quality before further analysis. These capabilities make qPCR invaluable for ensuring the accuracy and reliability of forensic investigations.

DETERMINING DEGRADATION LEVELS

Forensic samples are often exposed to environmental factors that can cause DNA degradation. Assessing the level of DNA degradation is critical for determining the quality and usability of the DNA sample for forensic analysis. qPCR can evaluate the degree of DNA degradation by amplifying different-sized DNA fragments. By comparing the amplification efficiency of short and long DNA targets, forensic scientists can infer the extent of degradation. Samples with higher degradation will show reduced amplification of longer fragments compared to shorter ones (Hedman et al., 2009). Understanding the degradation level of a DNA sample helps forensic scientists decide on the most suitable analysis methods. Highly degraded samples may require alternative approaches, such as using mini-STRs, which target shorter DNA regions, or next-generation sequencing methods that can analyze fragmented DNA more effectively.

ASSESSING SAMPLE QUALITY

Before proceeding with detailed forensic analysis, it is essential to assess the quality of the DNA sample. qPCR provides a means to evaluate the integrity and purity of the DNA extracted from forensic samples.

QUALITY CONTROL

qPCR can detect inhibitors that may be present in forensic samples and could potentially affect the efficiency of PCR amplification. Inhibitors can originate from various sources, such as soil, clothing, or bodily fluids. By including internal controls in the qPCR assays, forensic scientists can identify the presence of inhibitors and take corrective actions, such as further purifying the DNA sample (Butler, 2012).

DNA INTEGRITY

qPCR can also assess the integrity of DNA by measuring the ratio of nuclear DNA to mitochondrial DNA. This ratio can provide insights into the condition of the sample, as mitochondrial DNA is often more resistant to degradation compared to nuclear DNA. Samples with high nuclear DNA degradation may still contain usable mitochondrial DNA, which can be used for forensic identification (Parson & Dür, 2007).

Applications in forensic investigations

CRIME SCENE ANALYSIS

qPCR is used to analyze DNA samples collected from crime scenes, including blood, semen, saliva, and hair. By quantifying the DNA and assessing its quality, forensic scientists can determine the best approach for further analysis, such as STR profiling or mitochondrial DNA analysis.

IDENTIFICATION OF HUMAN REMAINS

qPCR is crucial in the identification of human remains, particularly in cases of mass disasters or missing persons. The ability to quantify and assess the quality of DNA helps forensic scientists identify remains accurately, even when the DNA is highly degraded or present in small quantities (Butler, 2012).

qPCR is also used in the analysis of cold cases and historical samples, where DNA may be highly degraded due to prolonged exposure to environmental factors. The sensitivity of qPCR allows for the detection and analysis of low-quality DNA, providing new leads and potential resolutions to long-standing cases (Schrader et al., 2012).

Advantages of qPCR

High sensitivity and specificity: qPCR can detect and quantify minute amounts of DNA or RNA, making it suitable for applications requiring high sensitivity. Sequence-specific probes enhance specificity, reducing the risk of detecting non-specific products (Kubista et al., 2006).

QUANTITATIVE DATA

qPCR provides quantitative data on the amount of target nucleic acids in the sample. This information is vital for applications like viral load monitoring, gene expression analysis, and mutation quantification.

RAPID AND HIGH-THROUGHPUT

qPCR is a rapid technique with results typically available within a few hours. It is also suitable for high-throughput applications, allowing the simultaneous analysis of multiple samples (Kubista et al., 2006).

REAL-TIME MONITORING

The ability to monitor DNA amplification in real time reduces contamination risk and provides immediate feedback on the reaction's progress. This feature is particularly useful for applications requiring precise quantification (Wong & Medrano, 2005).

WIDE DYNAMIC RANGE

qPCR can quantify DNA over a wide dynamic range, from a few copies to millions of copies, essential for accurately measuring varying DNA levels in different samples.

Challenges and considerations

While qPCR offers numerous advantages, some challenges and considerations need to be addressed.

PRIMER AND PROBE DESIGN

The design of primers and probes is critical for the success of qPCR. Poorly designed primers or probes can result in non-specific amplification, reduced sensitivity, and inaccurate quantification. Careful optimization is required to ensure specificity and efficiency (Bustin et al., 2009).

STANDARDIZATION AND QUALITY CONTROL

Standardization of qPCR protocols and rigorous quality control measures are essential to ensure reproducibility and accuracy. Variations in sample preparation, reaction conditions, and data analysis can affect the results. The MIQE guidelines provide a framework for standardizing qPCR experiments (Bustin et al., 2009).

INHIBITION AND SAMPLE QUALITY

The presence of inhibitors in the sample can affect the efficiency of qPCR. Inhibitors can originate from the sample matrix, extraction reagents, or other sources. It is crucial to assess sample quality and remove inhibitors to ensure accurate quantification (Nolan et al., 2006).

DATA ANALYSIS AND INTERPRETATION

qPCR generates large amounts of data that require careful analysis and interpretation. The selection of appropriate baseline and threshold values, as well as the use of reference genes for normalization, are critical for obtaining reliable results (Bustin et al., 2009).

Nested PCR

Nested PCR is an advanced variant of the traditional PCR technique that enhances the specificity and sensitivity of DNA amplification. In the context of molecular diagnostics, Nested PCR is particularly valuable for detecting low-abundance targets, minimizing non-specific amplification, and improving the accuracy of diagnostic tests. This technique is extensively used in various applications, including pathogen detection, genetic testing, and cancer diagnostics.

Nested PCR involves two rounds of PCR amplification using two sets of primers. The first round of PCR amplifies a larger region of the DNA template using the outer primers (**Figure 3.6**). The product from this amplification serves as the template for the second round of PCR, which uses the inner (nested) primers. These inner primers bind within the sequence amplified by the outer primers, ensuring that only the specific target sequence is further amplified (Sambrook & Russell, 2001).

Steps in nested PCR

First round of PCR

The DNA template is subjected to the first round of PCR using outer primers.
The outer primers anneal to the target sequence, and the DNA polymerase synthesizes the DNA.
This round produces a mixture of specific and non-specific PCR products.

Second round of PCR

The product from the first round is used as the template for the second round.
Nested primers, which are specific to the internal region of the first round product, are used in this round.
The second round selectively amplifies only the specific target sequence, reducing non-specific amplification.
By using two sets of primers, Nested PCR significantly enhances the specificity of DNA amplification. The likelihood of non-specific amplification products being further amplified in the second round is minimized, resulting in a higher yield of the desired product (Sambrook & Russell, 2001).

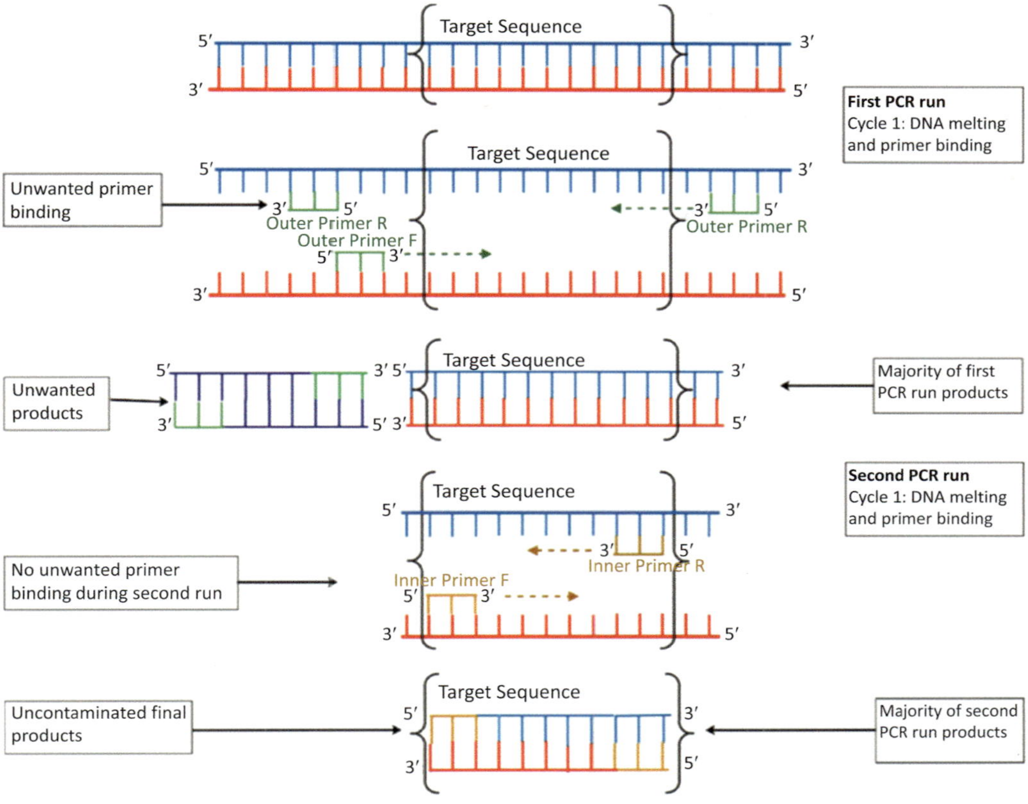

Figure 3.6 Step-by-step depiction of nested PCR showing two PCR runs with outer and inner primers to enhance specificity and reduce non-specific products.

Applications of nested PCR in molecular diagnostics

Nested PCR is a powerful and versatile tool in molecular diagnostics, offering enhanced specificity and sensitivity for detecting low-abundance targets and minimizing non-specific amplification. Its applications span a wide range of fields, including pathogen detection, genetic testing, cancer diagnostics, forensic science, and environmental monitoring.

Pathogen detection

Pathogen detection is one of the primary applications of Nested PCR, particularly for identifying bacterial, viral, and parasitic infections. The technique's enhanced specificity and sensitivity make it invaluable for diagnosing infectious diseases, even when the target DNA is present in low quantities or in samples with high background noise.

TUBERCULOSIS

TB, caused by *M. tuberculosis*, remains a significant global health challenge. Nested PCR is employed to detect *M. tuberculosis* in clinical samples, including sputum and tissue biopsies. The enhanced specificity and sensitivity of Nested PCR allow for the detection of TB even in samples with low bacterial load, improving diagnostic accuracy and enabling early intervention (Ou et al., 2008). This technique is particularly useful in cases where traditional culture methods are slow and less sensitive, providing rapid and reliable results.

MALARIA

Malaria diagnosis relies on detecting Plasmodium species in blood samples. Nested PCR offers superior sensitivity compared to conventional microscopy and can identify low levels of parasitemia, which is crucial for detecting asymptomatic carriers and monitoring treatment efficacy. By amplifying specific regions of the Plasmodium genome, Nested PCR can distinguish between different species of Plasmodium, facilitating accurate diagnosis and appropriate treatment (Snounou et al., 1993).

VIRAL INFECTIONS

Nested PCR is extensively used for the detection of various viral pathogens, including HIV, hepatitis B and C, and dengue virus. The high sensitivity allows for the detection of viral DNA or RNA even in samples with low viral load. For instance, Nested PCR is used to diagnose chronic hepatitis B and C infections by detecting viral DNA or RNA in blood samples, aiding in the management and treatment of these chronic diseases (Lok et al., 2001).

PARASITIC INFECTIONS

Besides malaria, Nested PCR is also utilized to detect other parasitic infections, such as toxoplasmosis and leishmaniasis. The technique's ability to amplify specific regions of the parasite's genome ensures accurate identification, even in complex biological samples. This is particularly important for diagnosing congenital infections, where early detection is crucial for effective treatment (Jones et al., 2001).

Genetic testing

In genetic testing, Nested PCR is invaluable for identifying specific mutations, polymorphisms, and genetic markers associated with various hereditary diseases. The high sensitivity allows for the detection of genetic variations even in low-quality DNA samples.

Cystic fibrosis: Cystic fibrosis is a hereditary disease caused by mutations in the CFTR gene. Nested PCR can detect common mutations in the CFTR gene, providing a reliable method for carrier screening and prenatal diagnosis. By amplifying the regions containing these mutations, Nested PCR ensures accurate detection, facilitating early intervention and management (Costes et al., 1993).

Thalassemia: Thalassemia is a genetic blood disorder caused by mutations in the HBB gene. Nested PCR is used to identify these mutations, allowing for the diagnosis and carrier screening of thalassemia. The technique's high sensitivity ensures the detection of mutations even in degraded DNA samples, making it a valuable tool for genetic testing in regions where thalassemia is prevalent (Newton et al., 1989).

Sickle cell anemia: Sickle cell anemia is another genetic disorder caused by a single nucleotide mutation in the HBB gene. Nested PCR can detect this mutation with high accuracy, enabling early diagnosis and management. The ability to amplify specific regions of the genome ensures that the mutation is accurately identified, even in samples with low DNA concentration (Miller et al., 1988).

Cancer diagnostics

In oncology, Nested PCR is employed to detect genetic mutations and chromosomal translocations associated with various cancers. The technique's high sensitivity allows for the detection of MRD after treatment, providing valuable information for monitoring disease progression and treatment response.

LEUKEMIA

Nested PCR is used to detect the BCR-ABL1 fusion gene in CML. The high sensitivity of Nested PCR enables the detection of MRD, providing valuable information for monitoring disease progression and assessing treatment efficacy. By amplifying specific regions of the BCR-ABL1 fusion gene, Nested PCR ensures accurate detection, even in cases with low levels of residual disease (Goh et al., 2011).

LYMPHOMA

In lymphoma diagnosis, Nested PCR can identify clonality by amplifying rearranged immunoglobulin or T-cell receptor genes. This technique helps distinguish between malignant and benign lymphoproliferative disorders, facilitating accurate diagnosis and appropriate treatment. The ability to detect specific genetic markers associated with lymphoma ensures that the disease is accurately characterized, guiding treatment decisions (Wan et al., 1999).

BREAST CANCER

Nested PCR is also used to detect mutations in the BRCA1 and BRCA2 genes associated with hereditary breast and ovarian cancers. The high sensitivity allows for the detection of these mutations even in samples with low DNA concentration, providing valuable information for risk assessment and management. By identifying individuals at high risk of developing cancer, Nested PCR facilitates early intervention and preventive measures (Ford et al., 1998).

Forensic science

Nested PCR is used in forensic science to analyze degraded or low-quantity DNA samples from crime scenes. The technique enhances the ability to obtain reliable DNA profiles from challenging samples, such as skeletal remains or trace biological evidence.

Forensic scientists use Nested PCR to analyze DNA samples collected from crime scenes, including blood, semen, saliva, and hair. The high sensitivity ensures that even small amounts of DNA can be detected and amplified, providing reliable evidence for criminal investigations. Nested PCR is particularly useful for analyzing degraded samples, where traditional PCR methods may fail (Butler, 2012).

In forensic anthropology, Nested PCR is used to identify human remains, particularly in cases of mass disasters or missing persons. The ability to amplify specific regions of the genome ensures accurate identification, even in highly degraded samples. By comparing the DNA profiles obtained from the remains with reference samples, forensic scientists can accurately identify individuals, providing closure for families and aiding in legal proceedings (Higgins et al., 2011).

Nested PCR is also applied in the analysis of cold cases, where DNA evidence may be highly degraded due to prolonged exposure to environmental factors. The high sensitivity allows for the detection and amplification of low-quality DNA, providing new leads and potential resolutions to long-standing cases. By reanalyzing old evidence with Nested PCR, forensic scientists can uncover new information that was previously undetectable (Schrader et al., 2012).

Multiplex PCR

Multiplex PCR is an advanced variant of the traditional PCR technique that allows for the simultaneous amplification of multiple target sequences in a single reaction. This method employs multiple sets of primers, each specific to a different target sequence, enabling the detection and quantification of several nucleic acid sequences simultaneously. The ability to amplify multiple targets in one reaction makes Multiplex PCR an efficient, cost-effective, and powerful tool in molecular diagnostics. This comprehensive analysis explores the extensive applications of Multiplex PCR in various fields of molecular diagnostics, highlighting its impact and relevance.

Multiplex PCR involves the simultaneous amplification of multiple target sequences in a single PCR reaction. This is achieved by including multiple pairs of primers, each specific to a different target sequence, in the reaction mix. The key to successful Multiplex PCR is the careful design and optimization of primers to ensure that they work efficiently together without interfering with each other.

Steps in multiplex PCR

Primer design

The design of primers is crucial for the success of Multiplex PCR. Primers must be specific to their target sequences and have similar melting temperatures (Tm) to ensure uniform amplification. Primer interactions, such as dimer formation and cross-hybridization, must be minimized through careful design and in silico analysis (Henegariu et al., 1997).

Optimization

Multiplex PCR requires optimization of reaction conditions, including $MgCl_2$ concentration, annealing temperature, and cycling parameters, to achieve efficient and specific amplification of all target sequences. This may involve testing different primer concentrations and reaction components (Elnifro et al., 2000).

Amplification

The PCR reaction is carried out in a thermal cycler, with multiple cycles of denaturation, annealing, and extension. Each primer pair amplifies its specific target sequence, resulting in the simultaneous generation of multiple PCR products. The amplified products can be detected and analyzed using various techniques, such as gel electrophoresis, capillary electrophoresis, or real-time PCR (Markoulatos et al., 2002).

Applications of multiplex PCR in molecular diagnostics

Multiplex PCR is an advanced variant of the PCR that allows simultaneous amplification of multiple target sequences in a single reaction. The ability to amplify multiple targets at once makes it highly efficient and cost-effective, leading to widespread applications in molecular diagnostics. The following sections elaborate on its use in various fields, including pathogen detection, genetic testing, cancer diagnostics, forensic science, and environmental and food safety.

Pathogen detection

RESPIRATORY INFECTIONS

Multiplex PCR is extensively used to diagnose respiratory infections by detecting multiple pathogens simultaneously. For example, it can identify influenza viruses, respiratory syncytial virus, and various bacterial pathogens in a single test. This capability is crucial during flu seasons and pandemics, where rapid and accurate diagnosis of co-infections can significantly influence patient management and treatment strategies (Mahony, 2008).

GASTROINTESTINAL INFECTIONS

The technique is also applied in diagnosing gastrointestinal infections by detecting multiple enteric pathogens, including bacteria (Salmonella, Shigella), viruses (norovirus, rotavirus), and parasites (Giardia, Cryptosporidium). Multiplex PCR reduces the time and cost of diagnosis by providing comprehensive results in a single test, improving patient outcomes through timely and appropriate treatment (Liu et al., 2012).

SEXUALLY TRANSMITTED INFECTIONS

Multiplex PCR can simultaneously detect multiple sexually transmitted infections (STIs), such as *Chlamydia trachomatis*, *Neisseria gonorrhoeae*, and *Trichomonas vaginalis*. This simultaneous detection ensures that co-infections are identified and treated appropriately, enhancing patient care and reducing the spread of infections (Chernesky et al., 2014).

Genetic testing

CYSTIC FIBROSIS

Multiplex PCR is used to detect common mutations in the CFTR gene associated with cystic fibrosis. By amplifying multiple regions of the gene in a single reaction, this technique provides a comprehensive screening method for carrier detection and prenatal diagnosis. This application is particularly valuable for identifying asymptomatic carriers and providing early intervention for affected individuals (Grody et al., 2001).

THALASSEMIA AND SICKLE CELL ANEMIA

Multiplex PCR can identify multiple mutations in the HBB gene, which are responsible for thalassemia and sickle cell anemia. The ability to detect various mutations simultaneously makes this technique highly efficient for genetic screening and diagnosis, particularly in populations with diverse genetic backgrounds (Dode & Guibaud, 1996).

GENETIC PANELS

Multiplex PCR is utilized in genetic panels designed to screen for multiple hereditary diseases. These panels can detect a wide range of genetic mutations associated with conditions such as hereditary cancers, cardiovascular diseases, and metabolic disorders. The simultaneous analysis of multiple targets enhances the efficiency and comprehensiveness of genetic testing, enabling early diagnosis and preventive healthcare (Rehm et al., 2013).

Cancer diagnostics

LEUKEMIA

In leukemia diagnosis, Multiplex PCR can detect multiple fusion genes and mutations. For instance, it can identify the BCR-ABL1 fusion gene in CML and mutations in the NPM1 and FLT3 genes in acute myeloid leukemia. This comprehensive approach provides critical information for diagnosis, prognosis, and treatment planning, enhancing personalized medicine strategies (Mrozek et al., 2009).

LYMPHOMA

Multiplex PCR can amplify rearranged immunoglobulin and T-cell receptor genes to identify clonality in lymphomas. This helps differentiate between malignant and benign lymphoproliferative disorders, facilitating accurate diagnosis and appropriate treatment (van Dongen et al., 2003).

SOLID TUMORS

The technique is also used to detect mutations in oncogenes and tumor suppressor genes in solid tumors, such as EGFR, KRAS, and BRAF mutations in lung and colorectal cancers. By providing comprehensive molecular profiling of tumors, Multiplex PCR aids in guiding targeted therapy and personalized treatment plans, improving patient outcomes (Ladanyi et al., 2008).

Forensic science

Multiplex PCR is a valuable tool in forensic science for analyzing DNA samples from crime scenes. The technique enhances the ability to obtain comprehensive DNA profiles, even from degraded or low-quantity samples.

CRIME SCENE ANALYSIS

Forensic scientists use Multiplex PCR to analyze DNA samples collected from crime scenes, including blood, semen, saliva, and hair. By amplifying multiple short tandem repeat (STR) loci simultaneously, Multiplex PCR provides detailed DNA profiles that can be used for individual identification, relationship testing, and criminal investigations (Butler, 2006).

IDENTIFICATION OF HUMAN REMAINS

In forensic anthropology, Multiplex PCR is used to identify human remains, particularly in cases of mass disasters or missing persons. The ability to amplify multiple genetic markers ensures accurate identification, even in highly degraded samples. By comparing the DNA profiles obtained from the remains with reference samples, forensic scientists can accurately identify individuals, providing closure for families and aiding in legal proceedings (Sijen, 2015).

Touchdown PCR

Touchdown PCR (Td-PCR) is a variant of the traditional PCR designed to enhance the specificity and yield of DNA amplification. This technique involves systematically decreasing the annealing temperature in successive cycles, allowing primers to bind more specifically to the target sequences. Touchdown PCR is particularly useful in molecular diagnostics, where precision and sensitivity are crucial. The following sections elaborate on the principle, mechanism, and applications of Touchdown PCR in various fields of molecular diagnostics.

Principle and mechanism

The principle of Touchdown PCR involves the gradual reduction of the annealing temperature in a stepwise manner during the initial cycles of PCR. This approach increases the stringency of primer binding during the early cycles, reducing non-specific binding and enhancing the specificity of the amplification.

Steps in touchdown PCR

Initial high-stringency annealing

The PCR starts with an annealing temperature higher than the optimal Tm (melting temperature) of the primers. This high temperature ensures that only the most specific primer-template pairs will anneal.

Gradual decrease in temperature

The annealing temperature is progressively reduced in successive cycles. Each step lowers the temperature closer to the primer's optimal Tm, allowing for increased primer binding.

Final annealing at optimal Tm

After the touchdown phase, the annealing temperature is set at the optimal Tm for the remaining cycles. This step ensures efficient amplification of the desired product.

By initially using a high annealing temperature and gradually decreasing it, Touchdown PCR minimizes non-specific amplification and enhances the yield of the specific target sequence (Don et al., 1991).

Applications of touchdown PCR in molecular diagnostics

Td-PCR is an advanced modification of the standard PCR technique that enhances the specificity and yield of DNA amplification. This is achieved by progressively lowering the annealing temperature in successive cycles, allowing primers to bind more accurately to target sequences. In molecular diagnostics, Td-PCR is particularly valuable due to its high specificity and sensitivity. Td-PCR has several applications in various fields, including pathogen detection, genetic testing, cancer diagnostics, forensic science, and environmental and food safety.

Pathogen detection

Pathogen detection is a critical application of Touchdown PCR in molecular diagnostics. The high specificity and sensitivity make it especially useful for identifying viral, bacterial, and parasitic infections.

VIRAL INFECTIONS

Td-PCR is extensively used in the diagnosis of viral infections due to its ability to detect low levels of viral DNA or RNA. For instance, in the detection of human papillomavirus (HPV), a pathogen associated with cervical cancer, Td-PCR allows for the identification of specific HPV genotypes, which is crucial for disease management and prevention (Yoshikawa et al., 1991). Similarly, it is employed in the diagnosis of hepatitis B and C, where precise detection of viral genotypes can guide appropriate antiviral therapies (Chen et al., 2008).

BACTERIAL INFECTIONS

The technique is also vital in detecting bacterial pathogens such as *M. tuberculosis*, the causative agent of TB. TB diagnosis often requires high specificity due to the complex nature of clinical samples, which may contain numerous other bacteria and contaminants. Touchdown PCR's ability to enhance specificity through a controlled annealing process helps in accurately identifying TB infections, even in samples with low bacterial loads (Rijpkema et al., 1996). Similarly, in the diagnosis of Lyme disease, caused by *Borrelia burgdorferi*, Touchdown PCR improves the detection accuracy, which is crucial for early intervention and treatment (Schwaiger et al., 2001).

PARASITIC INFECTIONS

In parasitic disease diagnostics, Touchdown PCR is used to detect DNA from parasites such as Plasmodium species (malaria) and Leishmania species. These infections often present in low-parasitemia cases, where traditional diagnostic methods might fail. The enhanced sensitivity and specificity of Touchdown PCR enable the detection of these parasites in various clinical samples, facilitating timely and accurate diagnosis (Cheng et al., 2014). Touchdown PCR's high specificity and sensitivity in pathogen detection significantly improve diagnostic accuracy, leading to better disease management and patient outcomes.

Genetic testing

In genetic testing, Touchdown PCR is employed to identify mutations, polymorphisms, and genetic markers associated with various hereditary diseases. The technique's high specificity minimizes the risk of amplifying non-target sequences, ensuring accurate genetic analysis.

SINGLE NUCLEOTIDE POLYMORPHISMS

Touchdown PCR is particularly effective in detecting SNPs because it can differentiate between closely related sequences by optimizing primer binding through a gradual decrease in annealing temperature (Kwok et al., 1994). This precise detection is essential for personalized medicine, where SNP profiles can guide individualized treatment plans.

INHERITED DISEASES

Touchdown PCR is crucial in diagnosing inherited diseases such as cystic fibrosis, thalassemia, and sickle cell anemia. These conditions are caused by specific genetic mutations that can be accurately detected using Touchdown PCR. For example, in cystic fibrosis, mutations in the CFTR gene can be identified with high specificity, facilitating carrier screening and prenatal diagnosis (Grody et al., 2001). Similarly, in thalassemia and sickle cell anemia,

Touchdown PCR improves the detection of mutations in the HBB gene, allowing for accurate genetic screening and management (Dode & Guibaud, 1996).

GENOTYPING

Touchdown PCR is also used for genotyping various genetic markers, including microsatellites and minisatellites. This application is valuable in genetic mapping, population genetics, and association studies. By enhancing primer specificity, Touchdown PCR ensures accurate amplification of specific loci, which is critical for reliable genotyping (Chakraborty et al., 1993).

Cancer diagnostics

In oncology, Touchdown PCR is employed to detect genetic mutations and chromosomal translocations associated with various cancers. The high specificity and sensitivity are crucial for identifying low-frequency mutations and monitoring MRD.

MUTATION DETECTION

Touchdown PCR is used to detect mutations in oncogenes and tumor suppressor genes, such as KRAS, EGFR, and TP53. Accurate detection of these mutations is essential for guiding targeted therapy and personalized treatment plans. For instance, mutations in the EGFR gene can predict responsiveness to tyrosine kinase inhibitors in non-small cell lung cancer, making precise detection vital for effective treatment (Ladanyi et al., 2008).

CHROMOSOMAL TRANSLOCATIONS

The technique is also applied to identify chromosomal translocations in hematological malignancies, such as the BCR-ABL1 fusion gene in CML. The high specificity of Touchdown PCR ensures reliable detection of these translocations, aiding in diagnosis and monitoring disease progression. This is particularly important for detecting MRD, where low levels of residual cancer cells must be accurately identified to assess treatment success and prevent relapse (Cross et al., 1994).

CIRCULATING TUMOR DNA

Touchdown PCR can be used to detect ctDNA in blood samples, providing a non-invasive method for monitoring tumor dynamics and detecting early relapse. The sensitivity allows for the detection of low levels of ctDNA, which is crucial for early intervention and treatment adjustment. This application is particularly valuable in liquid biopsy approaches, where ctDNA analysis can provide real-time insights into tumor evolution and treatment response (Diehl et al., 2008).

FORENSIC SCIENCE

Touchdown PCR is a valuable tool in forensic science for analyzing DNA samples from crime scenes. The technique enhances the ability to obtain reliable DNA profiles from challenging samples, such as degraded or low-quantity DNA.

Forensic scientists use Touchdown PCR to analyze DNA samples collected from crime scenes, including blood, semen, saliva, and hair. The high specificity ensures that even small amounts of DNA can be accurately amplified and analyzed, providing reliable evidence for criminal investigations. This is particularly important in cases where DNA samples are limited or degraded, as Touchdown PCR can improve the likelihood of obtaining usable genetic profiles (Butler, 2005).

Identification of human remains: Touchdown PCR is used to identify human remains, particularly in cases of mass disasters or missing persons. The technique's ability to amplify specific genetic markers ensures accurate identification, even in highly degraded samples. This application provides closure for families and aids legal proceedings by confirming the identity of individuals through DNA profiling (Higgins et al., 2011).

CONCLUSION

PCR has established itself as a cornerstone technology in molecular diagnostics since its development in the 1980s. By enabling the exponential amplification of specific DNA sequences, PCR has revolutionized the field of molecular biology, allowing for the detection of even the smallest amounts of nucleic acids in various samples. The fundamental principles of PCR, including the cycles of denaturation, annealing, and extension, form the basis of its ability to amplify DNA rapidly and accurately. This chapter has explored the critical components required for PCR, such as template DNA, primers, DNA polymerase, dNTPs, buffer solution, and magnesium ions, which are essential for the successful execution of the technique. It has also delved into the numerous variants of PCR, including real-time PCR (qPCR), Nested PCR, Multiplex PCR, and Touchdown PCR, each tailored to enhance the sensitivity, specificity, and versatility of the standard PCR method. These variants have expanded the range of applications for PCR, making it an indispensable tool in various fields of molecular diagnostics.

Table 3.1 Comparison of PCR variants

PCR variant	*Principle*	*Advantages*	*Common applications*
Conventional PCR	Amplifies DNA through repeated cycles of denaturation, annealing, and extension	Simple, cost-effective, widely used	Basic genetic testing, cloning, pathogen detection
Real-time PCR (qPCR)	Monitors DNA amplification in real-time using fluorescent dyes or probes	Quantitative data, high sensitivity, and specificity	Quantification of gene expression, viral load measurement
Nested PCR	Two rounds of PCR with two sets of primers for enhanced specificity	High specificity, reduces non-specific amplification	Detection of low-abundance targets, pathogen detection, genetic testing
Multiplex PCR	Simultaneous amplification of multiple targets in a single reaction	Efficient, cost-effective, high-throughput	Pathogen detection, genetic testing, cancer diagnostics
Touchdown PCR	Gradual decrease in annealing temperature to enhance specificity	High specificity, reduces non-specific amplification	Pathogen detection, genetic testing, cancer diagnostics

Table 3.2 Applications of real-time PCR (qPCR) in molecular diagnostics

Application	*Description*	*Examples*
Infectious disease detection	Detects and quantifies viral, bacterial, and fungal pathogens	HIV viral load measurement, COVID-19 detection, TB diagnosis
Genetic testing and mutation detection	Identifies genetic mutations, SNPs, and chromosomal aberrations	Detection of BRCA1/2 mutations, screening for cystic fibrosis mutations, prenatal testing for Down syndrome
Cancer diagnostics	Detects and quantifies tumor DNA and specific genetic alterations	Monitoring minimal residual disease (MRD), detection of KRAS, EGFR, and BRAF mutations
Gene expression analysis	Quantifies mRNA levels to study gene expression patterns	Biomarker discovery, validation of therapeutic targets, studying disease mechanisms
Forensic science	Quantifies DNA for forensic analysis and identification	Determining DNA degradation levels, assessing sample quality, crime scene analysis

The extensive applications of PCR and its variants have been highlighted throughout the chapter, emphasizing their role in pathogen detection, genetic testing, cancer diagnostics, forensic science, and environmental and food safety. Real-time PCR (qPCR), for instance, has become a critical method for quantifying DNA and RNA in real-time, providing valuable data for disease diagnosis and monitoring. Nested PCR and Multiplex PCR have improved the detection of specific sequences in complex samples, enhancing diagnostic accuracy and efficiency. Despite the significant advancements and widespread adoption of PCR, challenges remain, such as the need for meticulous optimization of reaction conditions and the potential for contamination leading to false positives. Nonetheless, techniques innovations and improvements in PCR technology continue to address these challenges, further solidifying its role as a fundamental technique in molecular diagnostics.

REFERENCES

Becker, K., Roth, R., & Peters, G. (2004). Rapid and specific detection of MRSA in routine clinical specimens by real-time PCR. Clinical Microbiology and Infection, 10(7), 635–641.

Branford S, Yeung D. T, Parker W. T, Roberts N. D, Purins L, Braley J. A, Altamura H. K, Yeoman A. L, Georgievski J, Jamison B. A, Phillis S, Donaldson Z, Leong M, Fletcher L, Seymour J. F, Grigg A. P, Ross D. M, Hughes T. P. (2014). Prognosis for patients with CML and >10% BCR-ABL1 after 3 months of imatinib depends on the rate of BCR-ABL1 decline. Blood, 124(4), 511–518. doi: 10.1182/blood-2014-03-566323.

Bustin, S. A., Benes, V., Garson, J. A., Hellemans, J., Huggett, J., Kubista, M., ... Wittwer, C. T. (2009). The MIQE guidelines: Minimum information for publication of quantitative real-time PCR experiments. Clinical Chemistry, 55(4), 611–622.

Butler, J. M. (2006). Genetics and genomics of core short tandem repeat loci used in human identity testing. Journal of Forensic Sciences, 51(2), 253–265.

Butler, J. M. (2012). Advanced topics in forensic DNA typing: methodology. Academic Press.

Butler, J. M. (2005). Forensic DNA typing: Biology, technology, and genetics of STR Markers. Elsevier Academic Press.

Chakraborty, R., Srinivasan, M. R., & Andrade, M. D. (1993). Analysis of population structure using DNA fingerprint data. Human Biology, 65(5), 875–887.

Chen, J., Wan, Z., Li, R., Yu, J., & Wu, Y. (2008). Development of a real-time PCR assay for detection and quantification of hepatitis B virus genotypes A to G. Journal of Virological Methods, 154(1–2), 102–108.
Cheng, Q., Xu, H., Liu, X., & Liu, X. (2014). Detection of Plasmodium species by polymerase chain reaction. Journal of Clinical Microbiology, 52(9), 3478–3481.
Chernesky, M., Jang, D., Luinstra, K., Chong, S., Sellors, J., & Mahony, J. (2014). Use of a multiplex polymerase chain reaction assay to simultaneously detect *Chlamydia trachomatis*, *Neisseria gonorrhoeae*, and *Trichomonas vaginalis* in genital specimens. Journal of Clinical Microbiology, 42(8), 3040–3044.
Corman, V. M., Landt, O., Kaiser, M., Molenkamp, R., Meijer, A., Chu, D. K., ... Drosten, C. (2020). Detection of 2019 novel coronavirus (2019-nCoV) by real-time RT-PCR. Euro Surveillance, 25(3), 2000045.
Costes, B., Girodon, E., Girodon-Boulandet, E., Martin, J., Petrescu, L., & Goossens, M. (1993). A rapid and efficient prenatal diagnosis of cystic fibrosis by the analysis of the length of DNA fragments amplified by polymerase chain reaction. Human Genetics, 91(2), 141–144.
Cross, N. C., Hughes, T. P., Feng, L., O'Shea, P., & Goldman, J. M. (1993). Minimal residual disease after allogeneic bone marrow transplantation for chronic myeloid leukemia: Detection of the BCR-ABL fusion gene by polymerase chain reaction. Blood, 82(8), 2603–2608.
Cross, N. C., Hughes, T. P., Feng, L., O'Shea, P., & Goldman, J. M. (1994). Minimal residual disease after allogeneic bone marrow transplantation for chronic myeloid leukemia: Detection of the BCR-ABL fusion gene by polymerase chain reaction. Blood, 84(2), 528–532.
Diehl, F., Schmidt, K., Choti, M. A., Romans, K., Goodman, S., Li, M., ... Vogelstein, B. (2008). Circulating mutant DNA to assess tumor dynamics. Nature Medicine, 14(9), 985–990.
Dode, C., & Guibaud, P. (1996). Direct detection of point mutations without DNA extraction or amplification: Application to β-thalassemia. Nucleic Acids Research, 24(4), 833–834.
Don R. H, Cox P. T, Wainwright B. J, Baker K, Mattick J. S. (1991). 'Touchdown' PCR to circumvent spurious priming during gene amplification. Nucleic Acids Research, 19(14), 4008.
Elnifro E. M, Ashshi A. M, Cooper R. J, Klapper P. E. (2000) Multiplex PCR: Optimization and application in diagnostic virology. Clinical Microbiology Reviews, 13(4), 559–570.
Erlich, H. A. (1989). PCR Technology: Principles and Applications for DNA Amplification. Stockton Press.
Ford, D., Easton, D. F., Bishop, D. T., Narod, S. A., & Goldgar, D. E. (1998). Risks of cancer in BRCA1-mutation carriers. The Lancet, 351(9119), 1389–1392.
Goh, H. G., Lin, M., Fukushima, T., Saglio, G., Kim, D. W., & Branford, S. (2011). Early molecular response and female sex independently predict better molecular response in chronic myeloid leukemia patients: Results from the international randomized study of interferon vs STI571 (IRIS) study. Clinical Cancer Research, 17(5), 1188–1197.
Grody, W. W., Desnick, R. J., Baron, H., Beaudet, A. L., Doyle, D., & Watson, M. (2001). ACMG position statement on universal newborn screening for cystic fibrosis. Genetics in Medicine, 3(2), 149–154.
Hedman, J., Nordgaard, A., & Rasmusson, B. (2009). Quantification and degradation analysis of forensic DNA samples using the novel QIAxcel system. Forensic Science International: Genetics Supplement Series, 2(1), 146–148.
Heid, C. A., Stevens, J., Livak, K. J., & Williams, P. M. (1996). Real-time quantitative PCR. Genome Research, 6(10), 986–994.
Henegariu, O., Heerema, N. A., Dlouhy, S. R., Vance, G. H., & Vogt, P. H. (1997). Multiplex PCR: Critical parameters and step-by-step protocol. BioTechniques, 23(3), 504–511.
Higgins, D., Austin, J. J., & Hagelberg, E. (2011). Using ancient DNA in the identification of human remains from an 18th century slave cemetery in the Cape Colony, South Africa. Journal of Archaeological Science, 38(2), 374–386.
Holland, P. M., Abramson, R. D., Watson, R., & Gelfand, D. H. (1991). Detection of specific polymerase chain reaction product by utilizing the 5′ to 3′ exonuclease activity of *Thermus aquaticus* DNA polymerase. Proceedings of the National Academy of Sciences, 88(16), 7276–7280
Innis, M. A., & Gelfand, D. H. (1990). PCR Protocols: A Guide to Methods and Applications. Academic Press.
Jones, C. D., Lambros, C., & Lowell, G. H. (2001). Nested PCR detection of *Toxoplasma gondii* in fetal tissue. Journal of Clinical Microbiology, 39(12), 4583–4585.
Kubista, M., Andrade, J. M., Bengtsson, M., Forootan, A., Jonák, J., Lind, K., ... Zoric, N. (2006). The real-time polymerase chain reaction. Molecular Aspects of Medicine, 27(2–3), 95–125.

Kwok, P. Y., Carlson, C., Yager, T. D., Ankoudinova, I., Trainor, G. L., & Nickerson, D. A. (1994). High-throughput genotyping of single nucleotide polymorphisms by competitive allele-specific PCR and fluorescence-based reporter detection. Nucleic Acids Research, 22(23), 4559–4562.

Ladanyi, M., Pao, W., & Chaft, J. E. (2008). Therapeutic targeting of anaplastic lymphoma kinase in lung cancer: Orchestrating clinical trial and biomarker development. Clinical Cancer Research, 14(21), 6672–6678.

Liu, J., Gratz, J., Amour, C. D., Kibiki, G., Becker, S., Janaki, L., ... Houpt, E. R. (2012). A laboratory-developed TaqMan array card for simultaneous detection of 19 enteropathogens. Journal of Clinical Microbiology, 51(2), 492–494.

Lok, A. S., & McMahon, B. J. (2001). Chronic hepatitis B. Hepatology, 34(6), 1225–1241.

Mahony, J. B. (2008). Multiplex PCR tests for respiratory infections. Journal of Clinical Virology, 40(Supplement 1), S15–S19.

Markoulatos, P., Siafakas, N., & Moncany, M. (2002). Multiplex polymerase chain reaction: A practical approach. Journal of Clinical Laboratory Analysis, 16(1), 47–51.

Miller, S. A., Dykes, D. D., & Polesky, H. F. (1988). A simple salting out procedure for extracting DNA from human nucleated cells. Nucleic Acids Research, 16(3), 1215.

Mrozek, K., Marcucci, G., Paschka, P., Whitman, S. P., & Bloomfield, C. D. (2009). Clinical relevance of mutations and gene-expression changes in adult acute myeloid leukemia with normal cytogenetics: Are we ready for a prognostically prioritized molecular classification? Blood, 109(2), 431–448.

Mullis, K., Faloona, F., Scharf, S., Saiki, R., Horn, G., & Erlich, H. (1986). Specific enzymatic amplification of DNA in vitro: The polymerase chain reaction. Cold Spring Harbor Symposia on Quantitative Biology, 51, 263–273.

Newton, C. R., Graham, A., Heptinstall, L. E., Powell, S. J., Summers, C., Kalsheker, N., ... Smith, J. C. (1989). Analysis of any point mutation in DNA. The amplification refractory mutation system (ARMS). Nucleic Acids Research, 17(7), 2503–2516.

Nolan, T., Hands, R. E., & Bustin, S. A. (2006). Quantification of mRNA using real-time RT-PCR. Nature Protocols, 1(3), 1559–1582.

Ou, X., Li, Q., Xia, H., Pang, H., Wang, S., Zhao, B., ... Xu, H. (2008). Diagnostic performance of the combination of interferon-γ release assays and nested PCR for *Mycobacterium tuberculosis*. Clinical Microbiology and Infection, 14(10), 1073–1078.

Pao, W., & Ladanyi, M. (2007). Epidermal growth factor receptor mutations in lung cancer: Implications for targeted therapies. Annual Review of Pathology: Mechanisms of Disease, 2, 163–184.

Parson, W., & Dür, A. (2007). EMPOP—A forensic mtDNA database. Forensic Science International: Genetics, 1(2), 88–92.

Rehm, H. L., Bale, S. J., Bayrak-Toydemir, P., Berg, J. S., Brown, K. K., Deignan, J. L., ... Working Group of the American College of Medical Genetics and Genomics Laboratory Quality Assurance Committee. (2013). ACMG clinical laboratory standards for next-generation sequencing. Genetics in Medicine, 15(9), 733–747.

Rijpkema, S. G., Golubic, M., & Molkenboer, M. J. (1996). Detection of *Borrelia burgdorferi* in clinical samples by PCR using enzyme-linked immunosorbent assay for product detection. Journal of Clinical Microbiology, 34(4), 854–859.

Saiki, R. K., Gelfand, D. H., Stoffel, S., Scharf, S. J., Higuchi, R., Horn, G. T., ... Mullis, K. B. (1988). Primer-directed enzymatic amplification of DNA with a thermostable DNA polymerase. Science, 239(4839), 487–491.

Sambrook, J., & Russell, D. W. (2001). Molecular Cloning: A Laboratory Manual (3rd ed.). Cold Spring Harbor Laboratory Press.

Schrader, C., Schielke, A., Ellerbroek, L., & Johne, R. (2012). PCR inhibitors–occurrence, properties and removal. Journal of Applied Microbiology, 113(5), 1014–1026.

Schwaiger, M., Péter, O., Cassinotti, P., & Wittenberg, R. (2001). Routine PCR detection of *Borrelia burgdorferi* sensu lato in clinical samples in comparison with culture and serology. Journal of Clinical Microbiology, 39(1), 346–351.

Siddiqi, N., Das, R., Pathak, N., Banerjee, S., Ahmed, N., & Hasnain, S. E. (2012). *Mycobacterium tuberculosis* isolate with a distinct genomic identity overexpresses a tap-like efflux pump. Journal of Infection, 64(5), 572–578.

Sijen, T. (2015). Molecular approaches for forensic cell type identification: On mRNA, miRNA, DNA methylation and microbial markers. Forensic Science International: Genetics, 18, 21–32.

Snounou, G., Viriyakosol, S., Jarra, W., Thaithong, S., & Brown, K. N. (1993). Identification of the four human malaria parasite species in field samples by the polymerase chain reaction and detection of a high prevalence of mixed infections. Molecular and Biochemical Parasitology, 58(2), 283–292.

Tyagi, S., & Kramer, F. R. (1996). Molecular beacons: Probes that fluoresce upon hybridization. Nature Biotechnology, 14(3), 303–308.
van Dongen, J. J., Langerak, A. W., Brüggemann, M., Evans, P. A., Hummel, M., Lavender, F. L., ... Wolvers-Tettero, I. L. (2003). Design and standardization of PCR primers and protocols for detection of clonal immunoglobulin and T-cell receptor gene recombinations in suspect lymphoproliferations: Report of the BIOMED-2 Concerted Action BMH4-CT98-3936. Leukemia, 17(12), 2257–2317.
Vogelstein, B., Kinzler, K. W. (2013). Digital PCR. Proceedings of the National Academy of Sciences, 96(16), 9236–9241.
Wan H, Aggleton JP, Brown MW. Different contributions of the hippocampus and perirhinal cortex to recognition memory. J Neurosci. 1999, 19(3), 1142–1148. doi: 10.1523/JNEUROSCI.19-03-01142.1999.
Whitcombe, D., Theaker, J., Guy, S. P., Brown, T., & Little, S. (1999). Detection of PCR products using self-probing amplicons and fluorescence. Nature Biotechnology, 17(8), 804–807.
White, P. L., Perry, M. D., Barnes, R. A., Johnson, E. M., & Archer, D. (2011). Candidaemia: Identification of a specific subpopulation at risk of high mortality. Journal of Antimicrobial Chemotherapy, 66(6), 1537–1541.
Wong, M. L., & Medrano, J. F. (2005). Real-time PCR for mRNA quantitation. BioTechniques, 39(1), 75–85.
Yoshikawa, H., Kawana, T., Kitagawa, K., & Mizuno, M. (1991). Amplification and typing of multiple cervical cancer-associated human papillomavirus DNAs using a single pair of primers. Journal of General Virology, 72(6), 1413–1417.
Zhang, F., Gu, W., Hurles, M. E., & Lupski, J. R. (2009). Copy number variation in human health, disease, and evolution. Annual Review of Genomics and Human Genetics, 10, 451–481.

Chapter 4

CRISPR-based diagnostic technologies

Asma Massad Alenzi

Department of Biology, Faculty of Science, University of Tabuk, Tabuk, Saudi Arabia

Biodiversity Genomics Unit, Faculty of Science, University of Tabuk, Tabuk, Saudi Arabia

Waad Aljohani

Department of Basic Health Sciences, College of Applied Medical Sciences, Qassim University, Buraydah, Saudi Arabia

Fawzyah Albaldi

Department of biology, Faculty of Science, Albaha University, Saudi Arabia

Mohammed Tarique[*]

Almanac Life Science India Private Limited, New Delhi, India

INTRODUCTION

CRISPR, which stands for Clustered Regularly Interspaced Short Palindromic Repeats, together with the related Cas (CRISPR-associated) proteins, forms an adaptive immune system in bacteria and archaea (**Figure 4.1**). This system defends against viral infections by specifically identifying and cutting foreign DNA. The technique has been utilized and modified for diverse applications, such as genome editing and molecular diagnostics, because of its exceptional specificity and effectiveness. The advancement and improvement of CRISPR-Cas systems have fundamentally transformed genetic research and medicinal methodologies.

CLASSIFICATION AND STRUCTURE

The CRISPR-Cas systems are categorized into two main classes based on the structure and function of their effector complexes: Class 1 and Class 2. Each class encompasses multiple types and subtypes, highlighting the diversity and adaptability of these systems.

Class I CRISPR-Cas systems

Class 1 systems employ multi-protein complexes to carry out their activity. These loci, as described by Makarova et al. (2020), are more varied and include around 90% of all known CRISPR-Cas loci. These systems are categorized into three kinds (I, III, and IV) and 22 subtypes. Class 1 systems are characterized by their dependence on heteromeric protein complexes consisting of several Cas proteins.

As an example, the Type I system, which is the most prevalent in Class 1, consists of proteins such as Cas3. Cas3 plays a vital role in degrading target DNA by performing helicase and nuclease functions. There are nine subtypes (I-A to I-G) within Type I systems, and separate subtypes like I-F1, I-F2, and I-F3 are considered derivatives of Type I. The Cas3

* Corresponding author.

DOI: 10.1201/9781003489177-4

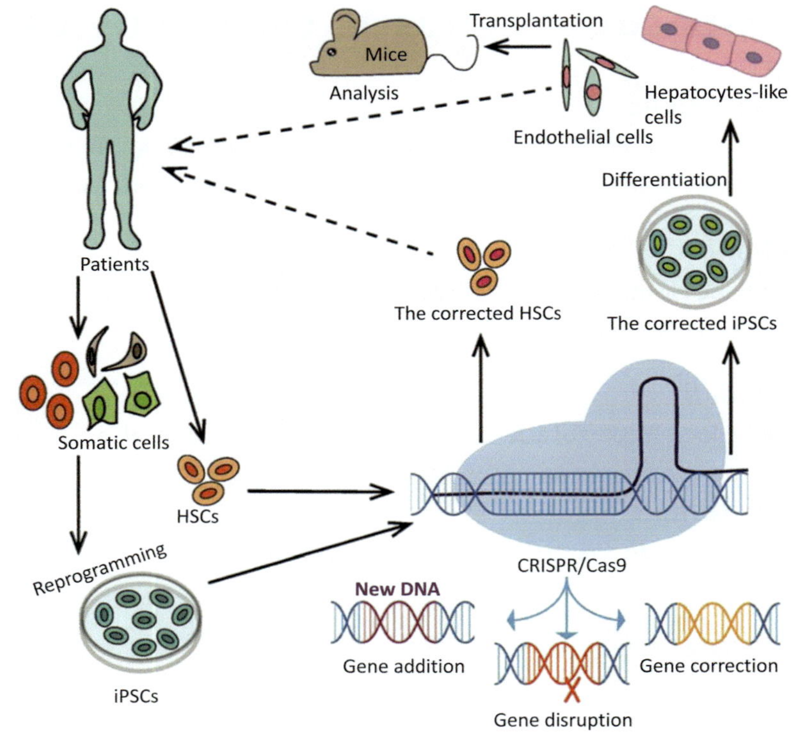

Figure 4.1 Diagram illustrating CRISPR/Cas9-mediated gene editing of patient-derived stem cells for therapeutic correction and transplantation.

protein functions as a helicase, which unwinds DNA-DNA or DNA-RNA double strands. It also possesses an HD family endonuclease domain that is responsible for cleaving the target DNA (Wright et al., 2016; Makarova et al., 2020).

Similarly, Type III systems, categorized under subtypes III-A to III-F, incorporate the Cas10 protein. Cas10 possesses a Palm domain, which is a kind of RNA recognition motif, and exhibits nucleic acid polymerase and cyclase capabilities. The effector complexes in these systems consist of a large subunit that is located in the Palm domain and is connected to an HD nuclease domain. This arrangement enables the cleavage of target DNA and its transcript by RNA-guided mechanisms, utilizing separate active sites within the complex (Makarova et al., 2020).

Type IV systems are relatively understudied; however, they generally lack the adaptive complex present in other kinds. These plasmid-like components frequently utilize plasmid targeting mechanisms to compete with other plasmids. These systems consist of proteins such as DinG family helicase or tiny alpha-helical proteins. These proteins rely on host-encoded restriction enzymes to break down their target molecules (Makarova et al., 2020).

Class 2 CRISPR-Cas systems

Class 2 systems are distinguished by the presence of single, multidomain effector proteins, which makes them less complex and more suitable for biotechnological uses. This class comprises three distinct kinds (II, V, and VI) and a total of 26 subtypes (Shmakov et al., 2017). The Type II system, which incorporates the extensively utilized Cas9 protein, is the most renowned Class 2 system.

Cas9, directed by a synthetic RNA molecule, may induce double-strand breaks (DSBs) in DNA at precise sites, enabling precise genome editing. This technique depends on a dual-RNA structure consisting of CRISPR RNA (crRNA) and trans-activating crRNA (tracrRNA) that directs the Cas9 enzyme to its specific DNA target. When Cas9 attaches to the target, it cuts the DNA. The DNA can then be fixed through either non-homologous end joining (NHEJ) or homology-directed repair (HDR), which enables accurate genetic alterations (Jinek et al., 2012; Doudna & Charpentier, 2014).

Additional prominent proteins within this category comprise Cas12 (Cpf1), which specifically recognizes DNA, and Cas13, which specifically recognizes RNA. Cas12a, for instance, cuts DNA in a staggered fashion, resulting in protruding ends that can be advantageous for particular forms of genetic manipulation. Conversely, Cas13 has been modified for the purpose of editing RNA and is especially valuable for identifying RNA viruses (Shmakov et al., 2017; Liu et al., 2023).

MECHANISM OF ACTION

The mechanism of action of the CRISPR-Cas system is a complex process that can be categorized into three primary stages: adaptation, expression, and interference. Every stage in this process is vital for the system's capacity to offer adaptive protection against viral infections in bacteria and archaea. Extensive research has focused on studying the exact molecular interactions and processes that occur during these stages. This research has greatly enhanced our understanding of CRISPR-Cas systems and has allowed us to modify them for genome editing and diagnostic purposes.

Adaptation

The adaptation phase, referred to as spacer acquisition, is the first stage of the CRISPR-Cas immune response. In this stage, the CRISPR system obtains more spacers, which are brief segments of viral DNA and incorporates them into the CRISPR array within the genetic material of the host. The Cas1 and Cas2 proteins collaborate to produce a complex that efficiently captures and inserts spacers between the CRISPR repeat sequences.

Spacer acquisition

The acquisition of spacers involves the recognition and cleavage of viral DNA. Cas1 and Cas2 proteins work together to identify protospacer sequences within the invading viral genome. These protospacers are typically flanked by short conserved sequences known as protospacer adjacent motifs (PAMs), which are crucial for the selection of spacer sequences. Once identified, the Cas1-Cas2 complex cleaves the viral DNA and integrates the resulting spacer into the CRISPR array, ensuring that the new spacer is flanked by repeat sequences (Wright et al., 2016; Makarova et al., 2020).

INTEGRATION INTO CRISPR ARRAY

The integration process consists of a sequence of coordinated actions in which the Cas1-Cas2 complex facilitates the insertion of the new spacer into the end of the CRISPR array that is close to the leader. This procedure guarantees that recently obtained spacers are situated in the immediate vicinity of the promoter region, enabling them to be transcribed with priority during subsequent infections. The spatial configuration of the system plays a vital role in its adaptive immunity, since spacers that are positioned closer to the promoter have a higher probability of being transcribed and converted into crRNAs (Wright et al., 2016).

Expression

The expression phase entails the conversion of the CRISPR array into a precursor CRISPR RNA (pre-crRNA) and its subsequent transformation into fully developed crRNAs by processing. This stage is crucial for producing the guide RNAs that will instruct the Cas proteins to their target regions.

TRANSCRIPTION OF CRISPR ARRAY

The CRISPR array is transcribed as a single elongated precursor CRISPR RNA (pre-crRNA), which consists of numerous repeat-spacer units. The transcription of the CRISPR array is initiated by a promoter situated upstream of it, guaranteeing the production of a single RNA molecule encompassing the complete array (Hille & Charpentier, 2016).

PROCESSING OF PRE-CRRNA

The pre-crRNA is subjected to processing to produce individual crRNAs, which consist of a spacer sequence and a repeat sequence. Within Class 1 systems, the Cas6 endoribonuclease is responsible for performing this processing. It cuts the pre-crRNA at the repeat sequences to release fully developed crRNAs. In Class 2 systems, the processing can be streamlined by a solitary protein, which simplifies the entire mechanism (Doudna & Charpentier, 2014).

FORMATION OF CRRNA-CAS COMPLEX

The mature crRNAs form complexes with the Cas proteins, creating the crRNA-Cas effector complexes that are essential for target recognition and cleavage. In Class 1 systems, these complexes are multi-protein assemblies, whereas in Class 2 systems, a single Cas protein, such as Cas9, binds to the crRNA to form the functional complex (Hille & Charpentier, 2016).

Interference

The interference phase is the final stage, where the crRNA-Cas complexes scan the cellular environment for complementary viral DNA or RNA sequences and cleave them, neutralizing the threat.

Target recognition

The crRNA, present in the crRNA-Cas complex, directs the complex to its target sequence by engaging in base-pairing interactions. The spacer sequence contained inside the crRNA exhibits complementarity to the target region present in the viral genome. Aside from base-pairing, the existence of a PAM sequence next to the target sequence is crucial for the identification of the target in numerous CRISPR systems. The Cas protein recognizes the PAM sequence, and it is necessary for the initiation of cleavage (Jinek et al., 2012).

CLEAVAGE OF TARGET DNA/RNA

After identifying the target, the Cas protein triggers the fragmentation of the target DNA or RNA. In Type II systems, the Cas9 protein triggers a DSB in the DNA, resulting in blunt ends that can be repaired by the cell's machinery. On the other hand, Cas12a (Cpf1) generates uneven incisions, resulting in adhesive ends. Cas13, which specifically recognizes and binds to RNA molecules, enzymatically cleaves single-stranded RNA molecules, rendering it well-suited for applications that include RNA manipulation (Shmakov et al., 2017).

Repair and immunity

The DSBs caused by Cas9 can be fixed via cellular mechanisms such as NHEJ or HDR. NHEJ is a repair mechanism that is prone to errors and frequently causes insertions or deletions at the location of a break, which ultimately disrupts the gene. HDR, however, is a mechanism for repairing genetic material with great accuracy and fidelity. It can be utilized for precise genome editing by supplying a donor template for the repair process (Cong et al., 2013).

CRISPR-Cas9 IN DISEASE DIAGNOSIS

CRISPR-Cas9 technology, known for its precision and efficiency in genome editing, has rapidly emerged as a transformative tool in disease diagnosis. The ability to accurately detect and characterize genetic mutations and pathogens makes CRISPR-Cas9 an invaluable asset in medical diagnostics. This section presents the mechanisms, applications, and case studies highlighting the use of CRISPR-Cas9 in diagnosing various diseases.

The detailed mechanisms by which CRISPR-Cas9 functions in the context of disease diagnostics, covering target recognition, cleavage, and reporter systems, along with recent enhancements and innovations in the field are also examined.

Mechanism

Target recognition and cleavage

The CRISPR-Cas9 system's diagnostic capability is predicated on its ability to recognize and bind to specific DNA sequences. This specificity is mediated by a guide RNA (gRNA) that directs the Cas9 protein to a complementary DNA sequence, where Cas9 induces a DSB.

Guide RNA (gRNA) design

The gRNA is composed of two constituents: the CRISPR RNA (crRNA) and the trans-activating crRNA (tracrRNA). The crRNA consists of a 20-nucleotide sequence that is complementary to the target DNA. This sequence directs the Cas9 protein to the precise site in the genome (Jinek et al., 2012). The tracrRNA functions as a framework that enhances the stability of the crRNA and aids in the creation of the gRNA-Cas9 complex. To streamline the method, these two components are frequently combined into a unified guide RNA for practical purposes (Doudna & Charpentier, 2014).

Cas9 protein activation

The Cas9 protein, specifically the form obtained from *Streptococcus pyogenes* (SpCas9), necessitates the existence of a PAM in close proximity to the target DNA sequence in order to attach and cleave it. The PAM sequence for SpCas9 is commonly

"NGG" with "N" representing any nucleotide. This requirement increases the precision of the system, guaranteeing that Cas9 only cuts DNA at the targeted locations (Jinek et al., 2012).

DNA BINDING AND CLEAVAGE

When the Cas9 protein attaches to the target DNA, it causes a DSB by using its two nuclease domains: RuvC and HNH. The RuvC domain specifically cleaves the DNA strand that is not complimentary, whereas the HNH domain specifically cleaves the DNA strand that is complementary. The detection and quantification of this DSB is crucial for diagnostic applications, since it can be accomplished utilizing many downstream approaches (Gasiunas et al., 2012).

REPORTER SYSTEMS AND SIGNAL DETECTION

The integration of reporter systems in CRISPR-Cas9 diagnostics is crucial for translating molecular events into detectable signals. These systems allow researchers to visualize and measure the occurrence of target DNA cleavage, which is essential for diagnosing various diseases accurately. There are several types of reporter systems used in conjunction with CRISPR-Cas9, including fluorescent, colorimetric, and electrochemical reporters (**Table 4.1**). Each type has its specific mechanism of action, advantages, and applications.

Fluorescent reporters

Fluorescent reporters are widely utilized due to their remarkable sensitivity and ability to provide immediate detection. These systems employ a DNA or RNA substrate that is labeled with a fluorescent dye and a quencher. The quencher suppresses fluorescence when it is intact by a mechanism known as fluorescence resonance energy transfer.

Table 4.1 Types of reporter systems in CRISPR-based diagnostics

Reporter type	*Mechanism*	*Benefits*	*Applications*
Fluorescent reporters	Fluorescence resonance energy transfer	High sensitivity, real-time detection	Pathogen detection, gene expression analysis, mutation identification
Colorimetric reporters	Enzyme-mediated color change upon cleavage	Simple, rapid, visible readout, no specialized equipment	Point-of-care testing, field diagnostics, low-resource settings
Electrochemical reporters	Changes in electrical properties upon target DNA/RNA binding and cleavage	High sensitivity, quantitative detection	Portable diagnostic devices, environmental monitoring, clinical diagnostics
Luminescent reporters	Emission of light upon cleavage of a luminescent substrate	High sensitivity, quantitative measurement	High-throughput screening, real-time monitorin

Once the Cas9 protein cuts, the dye and the quencher get separated, resulting in the creation of a visible fluorescent signal. This technology allows for continuous monitoring of the cleavage process, providing real-time data that can be quantified using fluorescence detection methods (Chen et al., 2018). Fluorescent reporters have an impressive degree of sensitivity, making them very suitable for detecting small amounts of target DNA. These qualities make them essential in several applications, including disease detection and the identification of genetic changes.

Colorimetric reporters

Colorimetric reporters provide a visible color difference when the target is detected, making them especially advantageous for point-of-care testing. These systems frequently employ enzymes that transform substrates lacking color into products with color upon activation by Cas9-induced cleavage. As an illustration, the splitting of a DNA substrate can liberate an enzyme that triggers a colorimetric reaction, resulting in a noticeable alteration in color. This technique is straightforward to execute and comprehend, without the need for specific equipment, making it appropriate for on-site diagnostics in settings with restricted resources (Gootenberg et al., 2017). The prompt outcomes offered by colorimetric tests are beneficial for circumstances where instant diagnostic information is crucial.

Electrochemical reporters

Electrochemical reporters are capable of detecting alterations in electrical signals when target DNA is cleaved. These technologies possess a high level of sensitivity and enable the precise measurement of DNA sequences. The electrode's surface is altered by attaching DNA probes that are complementary to the target sequence. After Cas9 binds to and cuts the target DNA, it is possible to quantify changes in the electrical parameters of the electrode, such as current or voltage. This method offers a swift and precise assessment of the existence of specific DNA (Koo et al., 2021). Electrochemical detection is highly advantageous for creating portable diagnostic devices that may be utilized in diverse circumstances, such as distant and resource-constrained areas.

The diagnostic capabilities of reporter systems have been significantly improved by recent breakthroughs in CRISPR technology. The advancement of CRISPR-Cas12 and Cas13 systems has broadened the scope of detectable nucleic acids. Cas12a specifically cleaves DNA by creating uneven cuts, whereas Cas13 specifically targets RNA, allowing for the identification of RNA viruses and other RNA-related biomarkers with exceptional accuracy and sensitivity (Chen et al., 2018; Freije et al., 2019). The multiplexing capabilities have been enhanced, enabling the concurrent identification of many targets in a single test. To enhance the throughput and efficiency of diagnostic testing, a method is employed where several guide RNAs are designed, each of which targets a distinct sequence (Ning et al., 2021).

The ongoing advancement of CRISPR-based diagnostics seeks to eradicate the necessity for nucleic acid amplification procedures, such as polymerase chain reaction (PCR), which have the potential to introduce errors and complicate the diagnostic process. Amplification-free detection methods leverage the high sensitivity of CRISPR systems to directly detect low-abundance targets, simplifying the workflow and reducing the time required for diagnostics (Chen et al., 2018).

MULTIPLEXING IN CRISPR DIAGNOSTICS

Multiplex CRISPR diagnostics represent a significant advancement in the field of molecular diagnostics, enabling the simultaneous detection of multiple genetic targets within a single assay. This approach leverages the versatility and precision of CRISPR-Cas systems, particularly the ability to design multiple guide RNAs (gRNAs), each specific to different DNA or RNA sequences. The development of multiplex CRISPR assays enhances the efficiency and throughput of diagnostic tests, making them invaluable in clinical settings where rapid and comprehensive diagnostics are essential.

DESIGN AND IMPLEMENTATION

The key to multiplex CRISPR diagnostics lies in the design of multiple gRNAs, each targeting a unique sequence within the sample. These gRNAs are co-delivered with the CRISPR-Cas machinery into the sample, where they simultaneously bind to their respective targets. The Cas proteins then induce DSBs or single-strand cuts at these specific sites, which can be detected using various reporter systems (Gootenberg et al., 2017; Ning et al., 2021).

ADVANTAGES OF MULTIPLEXING

1. By detecting multiple targets in a single reaction, multiplexing significantly increases the throughput of diagnostic assays. This is particularly beneficial in settings that require the screening of numerous pathogens or genetic mutations simultaneously.
2. Multiplex CRISPR diagnostics allow for the comprehensive screening of a wide array of genetic targets, providing a holistic view of the genetic landscape. This capability is crucial for identifying co-infections, genetic syndromes involving multiple mutations, or tracking the emergence of new pathogen variants.
3. Performing multiple tests in a single assay reduces the time and cost associated with diagnostics. This efficiency is especially important in resource-limited settings and during outbreaks, where rapid and affordable testing is critical (Chen et al., 2018).

Applications

Multiplex CRISPR diagnostics have been applied in various fields, including infectious disease diagnostics, genetic disorder screening, and cancer genomics. For instance, during the COVID-19 pandemic, multiplex CRISPR assays were developed to detect SARS-CoV-2 along with other respiratory pathogens, enabling comprehensive respiratory panel testing in a single assay (Broughton et al., 2020).

AMPLIFICATION-FREE DETECTION

Traditional nucleic acid detection methods, such as PCR, rely on amplification steps to increase the quantity of target DNA or RNA. While effective, these amplification steps can introduce errors, increase the complexity of the diagnostic process, and extend the time required to obtain results. CRISPR-based methods are being developed to detect low-abundance targets directly, without the need for amplification, leveraging the inherent sensitivity of CRISPR systems.

MECHANISM OF AMPLIFICATION-FREE DETECTION

CRISPR-based amplification-free detection methods utilize the high specificity and binding affinity of CRISPR-Cas proteins for their target sequences. These methods often employ CRISPR-Cas systems like Cas12 and Cas13, which can induce collateral cleavage of reporter molecules upon binding to their target, generating a detectable signal even at low target concentrations (Chen et al., 2018; Gootenberg et al., 2017).

ADVANTAGES

1. Eliminating the need for amplification simplifies the diagnostic workflow, reducing the number of steps and the potential for errors. This simplification is particularly beneficial for point-of-care and field diagnostics.
2. Amplification-free methods can provide rapid results, as they do not require the time-consuming thermal cycling steps associated with PCR. This rapid detection is crucial for timely clinical decision-making.
3. Amplification processes can increase the risk of contamination, leading to false positives. Amplification-free methods mitigate this risk, enhancing the reliability of the diagnostic results (Chen et al., 2018).

APPLICATIONS

Amplification-free CRISPR diagnostics have been utilized in several situations, encompassing the identification of viral infections, genetic mutations, and environmental surveillance (**Table 4.2**). CRISPR-Cas12-based assays have been created to directly identify SARS-CoV-2 RNA in clinical samples. These assays offer quick and precise diagnoses without the requirement of RNA extraction and amplification (Broughton et al., 2020).

Table 4.2 Applications of CRISPR-based diagnostics

Application	*Target*	*Method*	*Benefits*
Infectious disease detection	Pathogen DNA/RNA	Cas12/Cas13-based detection assays	Rapid, sensitive, and specific pathogen detection
Genetic disorder screening	Genetic mutations (e.g., CF, DMD)	Cas9-based mutation detection	Accurate identification of genetic mutations
Cancer diagnostics	Somatic mutations (e.g., EGFR, TP53)	Liquid biopsy with CRISPR-Cas9	Non-invasive, real-time monitoring of tumor dynamics
Prenatal testing	Fetal DNA in maternal blood	CRISPR-Cas9 targeting specific chromosomal abnormalities	Early detection of chromosomal abnormalities and single-gene disorders

APPLICATIONS IN DISEASE DIAGNOSIS

Amplification-free CRISPR diagnostics have been employed in various scenarios, including the detection of viral infections, genetic alterations, and environmental monitoring. CRISPR-Cas12 assays have been developed to directly detect SARS-CoV-2 RNA in clinical samples. These assays provide rapid and accurate diagnosis without the need for RNA extraction and amplification (Broughton et al., 2020).

GENETIC DISORDERS

The CRISPR-Cas9 technology has had a profound impact on the detection of genetic disorders because of its versatility and precision. It has unparalleled capabilities for detecting and analyzing genomic abnormalities. This section provides a detailed explanation of the utilization of CRISPR-Cas9 in the detection of genetic abnormalities, with a specific emphasis on the recognition of genetic mutations, prenatal screening, and carrier screening.

IDENTIFICATION OF GENETIC MUTATIONS

The precise targeting and cleaving of certain DNA sequences by CRISPR-Cas9 has rendered it an indispensable tool for discovering genetic alterations linked to different inherited illnesses. Genetic mutations frequently appear as single nucleotide polymorphisms, insertions, deletions, or more intricate genomic rearrangements. Identifying these mutations is essential for the diagnosis of illnesses such as cystic fibrosis (CF), sickle cell anemia, and Duchenne muscular dystrophy (DMD).

CYSTIC FIBROSIS

CF is a genetic disorder resulting from mutations in the CFTR gene, with the ΔF508 variant being the most prevalent. Conventional diagnostic techniques for CF, like sweat tests and genetic testing, may be time-consuming and have lower sensitivity. CRISPR-Cas9 provides a fast and accurate alternative for identifying CFTR mutations. Researchers can utilize CRISPR-Cas9 to create DSBs at the mutation site by developing guide RNAs (gRNAs) that precisely target the ΔF508 region of the CFTR gene. The following examination of cleavage patterns provides a definitive indicator of whether the mutation is present or absent (Kosicki et al., 2018).

SICKLE CELL ANEMIA

Sickle cell anemia is caused by a single nucleotide mutation in the HBB gene, leading to the production of abnormal hemoglobin. CRISPR-Cas9 can be employed to detect this specific mutation by using gRNAs that target the mutated region. The high specificity of CRISPR-Cas9 allows for the differentiation between the normal and mutant alleles, enabling accurate diagnosis. Furthermore, CRISPR-based assays can be adapted for high-throughput screening, facilitating the diagnosis of sickle cell anemia in large populations (Dever et al., 2016).

DUCHENNE MUSCULAR DYSTROPHY

DMD is a genetic disorder that is inherited through the X chromosome and is caused by mutations in the dystrophin gene. The extensive length of the dystrophin gene and the diverse range of mutations make its diagnosis more complex. The CRISPR-Cas9 system offers a flexible method for identifying deletions, duplications, and point mutations in the dystrophin gene. CRISPR-Cas9 can be employed to detect the precise alterations in a patient's genome by creating several gRNAs that target various sections of the gene. This technique not only improves the accuracy of diagnosis but also assists in the formulation of individualized treatment plans (Long et al., 2018).

TECHNICAL ADVANCES IN MUTATION DETECTION

Recent developments in CRISPR technology have enhanced its effectiveness in detecting mutations. Methods such as CRISPR-mediated HDR and base editing have enhanced the accuracy of mutation detection. HDR facilitates the incorporation of a repair template that rectifies the mutation during the DNA repair mechanism, whereas base editing permits the direct conversion of one nucleotide to another without causing DSBs. These advancements improve the ability of CRISPR-based diagnostics to accurately and precisely detect specific targets, resulting in increased reliability and efficiency (Komor et al., 2016).

PRENATAL SCREENING

Prenatal screening is a critical application of CRISPR-Cas9, providing early detection of genetic abnormalities in fetuses. Non-invasive prenatal testing (NIPT) has gained popularity for its ability to analyze cell-free fetal DNA (cffDNA) circulating in maternal blood. CRISPR-Cas9 enhances NIPT by enabling precise targeting and analysis of fetal DNA sequences.

EARLY DETECTION OF CHROMOSOMAL ABNORMALITIES

CRISPR-Cas9 can be used to identify chromosomal abnormalities like Down syndrome (trisomy 21), Edwards syndrome (trisomy 18), and Patau syndrome (trisomy 13). CRISPR-Cas9 can create DSBs at precise locations on the afflicted chromosomes by using designed gRNAs. An increased number of these chromosomes in the cffDNA leads to unique patterns of cleavage, which can be examined to verify the existence of chromosomal abnormalities. This technology offers a precise and non-intrusive strategy for the early identification of these diseases (Benn et al., 2015).

DETECTION OF SINGLE-GENE DISORDERS

CRISPR-Cas9 can be used to diagnose single-gene diseases, such as thalassemia and CF, before birth. CRISPR-Cas9 can induce accurate incisions at the mutation locations by selectively targeting the exact mutations accountable for these illnesses. By sequencing the cleaved DNA fragments that follow, it becomes possible to identify these mutations, which in turn allows for early diagnosis and action. This tool is especially beneficial for families that have a documented history of single-gene illnesses, since it enables them to make well-informed decisions and arrange for early treatment interventions (Wang et al., 2023).

ETHICAL CONSIDERATIONS

CRISPR-Cas9 presents notable benefits for prenatal screening, but it also gives rise to ethical concerns. It is crucial to thoroughly assess the possibility of unintended impacts and the precision of NIPT in order to guarantee the dependability of the outcomes. Furthermore, the ethical considerations surrounding the early identification of genetic conditions and the potential for choosing to end pregnancies based on genetic information require careful ethical discussions and informed consent procedures. It is necessary to create regulatory frameworks in order to tackle these difficulties and guarantee the conscientious use of CRISPR-Cas9 in prenatal diagnostics (Benn et al., 2015).

CARRIER SCREENING

Carrier screening is essential for identifying individuals who carry recessive genetic mutations that could be passed on to their offspring. CRISPR-Cas9 enhances carrier screening by providing accurate and high-throughput detection of carrier status for various genetic disorders.

High-throughput screening

CRISPR-Cas9 enables efficient and rapid screening of carrier status in huge populations. CRISPR-Cas9 can efficiently determine carrier status by utilizing gRNAs that are designed to target known mutations in genes linked to recessive diseases. This method is especially beneficial for screening programs that target the entire community with the goal of detecting individuals who carry common recessive illnesses like CF, Tay-Sachs disease, and spinal muscular atrophy. High-throughput CRISPR-Cas9 assays have the capability to handle numerous samples concurrently, hence diminishing the time and expenses linked to conventional screening techniques (Kosicki et al., 2018).

Comprehensive genetic screening

CRISPR-Cas9 facilitates extensive genetic screening by enabling the concurrent identification of numerous mutations in a single experiment. Multiplex CRISPR diagnostics employ numerous gRNAs that target distinct mutations, enabling the identification of carriers of diverse genetic illnesses in a single test. This feature is especially valuable in clinical environments where extensive genomic data is necessary for counseling and decision-making. As an illustration, couples who are planning to have children can undergo carrier screening to evaluate their likelihood of transmitting genetic problems to their offspring (Ning et al., 2021).

Personalized medicine

CRISPR-Cas9-based carrier screening also facilitates the advancement of personalized therapy. Through precise identification of carriers with specific mutations, clinicians can offer tailored guidance for family planning, genetic counseling, and risk evaluation. This strategy enables individuals to have access to the necessary knowledge to make well-informed decisions regarding their reproductive health and to effectively handle the potential risks of genetic abnormalities within their families (Dever et al., 2016).

Overall, the utilization of CRISPR-Cas9 in the identification of genetic diseases is extensive and revolutionary. The precision and efficiency of this technology allow for the precise identification of genetic mutations, early detection of chromosomal abnormalities, and thorough carrier screening. These features greatly improve the diagnostic process by offering crucial information for early intervention, tailored treatment, and well-informed decision-making. The ongoing advancement of CRISPR technology is expected to have a significant impact on the diagnosis and treatment of genetic abnormalities, presenting novel prospects for enhancing patient care and public health.

DIAGNOSIS OF INFECTIOUS DISEASES

CRISPR-Cas9 technology has proven to be a powerful tool for diagnosing infectious diseases, offering rapid, sensitive, and specific detection of pathogens. This capability is vital for managing infectious diseases, improving patient outcomes, and controlling outbreaks. The application of CRISPR-Cas9 in infectious disease diagnostics spans various pathogens, including viruses, bacteria, and even fungi.

VIRAL DETECTION

Viruses present a significant challenge for diagnostic technologies due to their rapid mutation rates and diverse genetic makeup. CRISPR-Cas9 has shown great promise in overcoming these challenges by enabling precise and specific detection of viral genomes.

HIV detection

CRISPR-Cas9 is widely recognized for its significant role in identifying HIV-1. Conventional techniques for HIV-1 detection, such as enzyme-linked immunosorbent assay (ELISA) and PCR, can be laborious and necessitate sophisticated equipment. CRISPR-Cas9-based assays provide a more expedient option by specifically targeting HIV-1 DNA sequences. Researchers can utilize CRISPR-Cas9 to create DSBs at specific places in the HIV-1 genome by developing gRNAs that target those areas. Reporter systems can be employed to detect subsequent cleavage events, offering a prompt and highly sensitive approach to identify HIV-1 in clinical samples (Ning et al., 2021).

SARS-CoV-2 detection

The COVID-19 pandemic emphasized the necessity for prompt and precise diagnostic instruments. CRISPR-Cas9 and its derivatives, including CRISPR-Cas12 and CRISPR-Cas13, have been modified to identify SARS-CoV-2 RNA. These systems utilize the collateral cleavage activity of Cas12 and Cas13, which is triggered when they attach to their target RNA sequence. This activity can be combined with fluorescent or colorimetric reporters to generate a visible indication. Assays based on CRISPR-Cas12, such as DETECTR, and CRISPR-Cas13, such as SHERLOCK, have shown excellent sensitivity and specificity. They can rapidly detect the presence of SARS-CoV-2 in patient samples, typically within minutes (Gootenberg et al., 2017; Broughton et al., 2020).

Zika and dengue virus detection

CRISPR-Cas9 has also been employed to detect other RNA viruses such as Zika and dengue viruses. These viruses are significant public health concerns in tropical and subtropical regions. By targeting specific RNA sequences of these viruses, CRISPR-Cas9-based diagnostics can provide rapid and accurate detection, aiding in the timely management of outbreaks and reducing the burden of these diseases on healthcare systems (Freije et al., 2019).

BACTERIAL DETECTION

Bacterial infections, especially those caused by antibiotic-resistant strains, pose a severe threat to public health. CRISPR-Cas9 technology offers a robust solution for detecting bacterial pathogens and their resistance genes.

Detection of methicillin-resistant *Staphylococcus aureus*

Methicillin-resistant *S. aureus* (MRSA) is a prevalent source of nosocomial infections and is renowned for its resistance to methicillin and other antimicrobial agents. CRISPR-Cas9 can be employed for MRSA detection by specifically targeting the *mecA* gene, responsible for conferring methicillin resistance. CRISPR-Cas9 can detect the presence of MRSA in clinical samples by creating gRNAs that are tailored to the *mecA* gene. This approach offers a swift and precise substitute for conventional culture-based methods, which are laborious and necessitate specific laboratory conditions (Gootenberg et al., 2017).

Detection of tuberculosis

Tuberculosis (TB), caused by the bacteria *Mycobacterium tuberculosis*, continues to be a significant worldwide health problem. CRISPR-Cas9-based diagnostic methods have been created for the purpose of identifying *M. tuberculosis* DNA in samples of sputum. These tests can selectively focus on distinct genomic areas that are exclusive to *M. tuberculosis*, offering a quick and precise approach for detecting TB. The exceptional sensitivity of CRISPR-Cas9 enables the detection of minimal amounts of bacteria, a critical factor in the identification of tuberculosis at its early stages and the assessment of treatment effectiveness (Chen et al., 2018).

Multiplex detection of bacterial pathogens

The ability to detect multiple bacterial pathogens simultaneously is a significant advantage of CRISPR-Cas9 diagnostics. Multiplex assays can be designed to target multiple bacterial species in a single test, enabling comprehensive screening for co-infections. This capability is particularly useful in clinical settings where patients may present complex infections involving multiple pathogens. For example, multiplex CRISPR diagnostics can screen for respiratory pathogens such as *Streptococcus pneumoniae*, *Haemophilus influenzae*, and *Legionella pneumophila*, providing valuable information for accurate diagnosis and treatment (Ning et al., 2021).

ADVANCEMENTS AND FUTURE DIRECTIONS

Recent advancements in CRISPR technology have significantly improved its capacity for infectious disease diagnoses. CRISPR-Cas12 and Cas13 systems, which have the ability to specifically target single-stranded DNA and RNA, have expanded the range of applications for CRISPR-based diagnostics. These systems utilize the collateral cleavage activity of Cas12 and Cas13, which can be exploited to produce enhanced signals when the target is identified.

CRISPR-Cas12 and Cas13 systems

CRISPR-Cas12a (Cpf1) and CRISPR-Cas13 systems have been developed to detect DNA and RNA viruses with high sensitivity and specificity. Cas12a targets double-stranded DNA and creates staggered cuts, while Cas13 targets RNA and exhibits collateral cleavage activity that can degrade nearby RNA molecules, enhancing signal amplification. These properties make Cas12a and Cas13 ideal for developing point-of-care diagnostics that provide rapid results with minimal equipment (Chen et al., 2018; Freije et al., 2019).

Point-of-care diagnostics

Researchers are currently working on developing CRISPR-based point-of-care diagnostics, which will allow for quick testing to be conducted in non-traditional laboratory environments. These portable and intuitive gadgets can be utilized in distant or resource-constrained regions, delivering prompt and precise diagnosis. CRISPR-based paper strip tests, like pregnancy tests, have been created to identify viral RNA, as an illustration. These tests provide a visual display of results and may be conducted without requiring professional training, thereby making them easily accessible to a wide range of users (Broughton et al., 2020).

Integration with next-generation sequencing

Integrating CRISPR-Cas9 diagnostics with next-generation sequencing (NGS) can improve the detection and characterization of pathogens. This integration enables the concurrent detection of several infections and the examination of their genetic variability. The utilization of CRISPR-Cas9 allows for the enhancement of specific sequences from intricate samples, hence enhancing the sensitivity and precision of diagnostics based on NGS. This method is very useful for observing the development of infectious disorders and tracing the dissemination of genes that resist antimicrobial substances (Gootenberg et al., 2017).

CRISPR-Cas9 IN CANCER DIAGNOSTICS

CRISPR-Cas9 technology has demonstrated significant potential in identifying somatic mutations associated with various cancers. By targeting specific cancer-related genes, CRISPR-based diagnostics can detect mutations that drive tumor growth and progression. This capability is crucial for early diagnosis and personalized treatment planning, which are key to improving patient outcomes in oncology.

MECHANISM OF CRISPR-Cas9 IN CANCER DIAGNOSTICS

The CRISPR-Cas9 system's ability to precisely edit and detect DNA sequences underlies its potential in cancer diagnostics. The technology involves designing guide RNAs (gRNAs) that target specific sequences in cancer-related genes. When these gRNAs guide the Cas9 protein to the target site, Cas9 induces a DSB in the DNA, which can then be repaired and analyzed to confirm the presence of mutations.

TARGETING CANCER-RELATED GENES

Specific genes are frequently mutated in various cancers. For example, mutations in the TP53 gene are common in many types of cancer, including breast, lung, and colorectal cancers (Levine et al., 2019). Similarly, mutations in the EGFR gene are significant in non-small cell lung cancer (NSCLC) (Zhang et al., 2019). By designing gRNAs that target these specific mutations, CRISPR-Cas9 can be used to detect the presence of these mutations in tumor DNA.

DETECTION OF MUTATIONS

Once the CRISPR-Cas9 system introduces a DSB at the target site, various downstream techniques can be employed to detect and analyze the mutations. These techniques include sequencing the cleaved DNA fragments to confirm the presence of specific mutations. The high specificity of CRISPR-Cas9 ensures that only the intended DNA sequences are targeted, reducing the likelihood of false positives (Cong et al., 2013).

APPLICATIONS IN EARLY DIAGNOSIS AND MONITORING

CRISPR-Cas9's precision and efficiency make it an invaluable tool for early cancer diagnosis and monitoring. Early detection of cancer significantly increases the chances of successful treatment and improved survival rates.

Liquid biopsies

Liquid biopsies entail the examination of circulating tumor DNA (ctDNA) present in blood samples. CRISPR-Cas9 has the ability to improve the accuracy and precision of liquid biopsies by selectively targeting certain mutations in ctDNA. The non-invasive method described enables the timely identification of cancer and the continuous observation of tumor changes, offering crucial data for making treatment choices (Wan et al., 2017).

Liquid biopsies offer distinct benefits as they have the ability to capture the diverse characteristics of tumors and identify even small amounts of remaining disease. Conventional tissue biopsies offer a limited view of the tumor at a certain moment and location, potentially overlooking alterations found in other areas of the tumor or in metastatic regions. On the other hand, circulating tumor DNA (ctDNA) in the blood can offer a more thorough understanding of the genetic characteristics of the malignancy, detecting mutations from different tumor locations (Wan et al., 2017).

Monitoring treatment response

CRISPR-Cas9 can be utilized to assess the efficacy of cancer therapies by identifying any remaining illness or newly developed mutations that confer resistance. For instance, CRISPR-based tests can identify mutations in the EGFR gene that cause resistance to tyrosine kinase inhibitors (TKIs) in patients with NSCLC. The provided information allows medical professionals to modify treatment approaches and enhance the results for patients (Zhang et al., 2019).

Monitoring therapy response is essential for customizing medicines for specific patients. By identifying resistance mutations at an early stage, medical professionals can transition to alternate therapies prior to the advancement of the tumor. For example, in cases of breast cancer, mutations in the ESR1 gene can provide resistance to hormonal therapy. CRISPR-Cas9-based diagnostics have the ability to detect these changes in ctDNA, enabling prompt modifications in the treatment plan (O'Leary et al., 2018).

Personalized treatment planning

CRISPR-Cas9 can be employed to evaluate the effectiveness of cancer treatments by detecting any residual disease or newly acquired mutations that provide resistance. CRISPR-based testing can specifically detect mutations in the EGFR gene that lead to resistance against TKIs in individuals diagnosed with NSCLC. The information supplied enables medical practitioners to adapt treatment procedures and optimize outcomes for patients (Zhang et al., 2019).

Monitoring the response to therapy is crucial for tailoring medications to individual patients. Through the early detection of resistance mutations, healthcare providers can switch to alternative treatments before the tumor progresses. In instances of breast cancer, mutations in the ESR1 gene can confer resistance to hormone therapy. CRISPR-Cas9-based diagnostics can identify these alterations in ctDNA, allowing for timely adjustments in the treatment strategy (O'Leary et al., 2018).

CASE STUDIES AND RESEARCH ADVANCES

Several studies have demonstrated the effectiveness of CRISPR-Cas9 in detecting cancer-related mutations and guiding personalized treatment.

Lung cancer

A study utilized CRISPR-Cas9 to identify genetic abnormalities in the EGFR gene among patients diagnosed with NSCLC. The CRISPR-based technique effectively detected EGFR mutations in ctDNA extracted from blood samples, enabling non-invasive cancer diagnosis and the monitoring of therapy response. The implementation of CRISPR-Cas9 in liquid biopsy demonstrates its potential as a minimally invasive and highly precise technique for cancer diagnostics (Zhang et al., 2019).

EGFR mutations play a crucial role in driving NSCLC, and it is vital to discover them in order to guide treatment with TKIs. CRISPR-Cas9 diagnostics have the ability to distinguish between sensitizing mutations, which make cells more sensitive to TKIs, and resistance mutations, which suggest that alternative therapies are necessary. The accurate identification of this condition allows for more efficient development of treatment strategies and enhances the overall results for patients (Zhang et al., 2019).

Breast cancer

Studies have demonstrated that CRISPR-Cas9 has the capability to identify genetic alterations in the BRCA1 and BRCA2 genes among individuals with breast cancer. These mutations play a critical role in predicting the likelihood of developing breast and ovarian cancers and in guiding preventive and treatment approaches. CRISPR-based assays offer a very accurate and effective technique for detecting these mutations, hence supporting personalized medicine strategies (Levine et al., 2019).

BRCA1 and BRCA2 mutations greatly enhance the likelihood of developing breast and ovarian cancers. Identifying these genes is crucial for taking preventive actions such as prophylactic surgery and utilizing targeted treatments like PARP inhibitors. CRISPR-Cas9 diagnostics provide the ability to detect these alterations with great accuracy, facilitating prompt intervention and tailored therapy strategizing (Lord & Ashworth, 2016).

Colorectal cancer

Tumor growth and progression in colorectal cancer are frequently linked to mutations in the APC and KRAS genes. The application of CRISPR-Cas9 technology has facilitated the detection of these alterations in tumor samples, hence assisting in the early identification and individualized planning of treatment. CRISPR-based diagnostics enhance patient outcomes by offering precise genetic information, which in turn enables tailored medicines (Cong et al., 2013).

The occurrence of mutations in the APC gene is one of the initial stages in the formation of colorectal cancer, whereas mutations in the KRAS gene promote the advancement of tumors and make them resistant to specific treatments. CRISPR-Cas9 diagnostics can detect these mutations, enabling early identification and the choice of suitable targeted therapy, such as EGFR inhibitors for patients with KRAS wild-type malignancies (Amado et al., 2008).

FUTURE DIRECTIONS AND CHALLENGES

The future of CRISPR-Cas9 in cancer diagnostics looks promising, with ongoing research aimed at enhancing its accuracy, efficiency, and clinical applicability.

Enhancing specificity and reducing off-target effects

One of the primary challenges in CRISPR-Cas9 diagnostics is ensuring high specificity and minimizing off-target effects. Advances in gRNA design and Cas9 variants are being developed to enhance the precision of CRISPR-based assays, reducing the risk of false positives and negatives (Chen et al., 2018).

Off-target effects occur when CRISPR-Cas9 introduces DSBs at unintended sites in the genome. These off-target effects can lead to false-positive results and potentially harmful genomic alterations. To address this issue, researchers are developing high-fidelity Cas9 variants with reduced off-target activity and optimizing gRNA design algorithms to improve target specificity (Chen et al., 2018).

Integration with next-generation sequencing

Combining CRISPR-Cas9 with NGS can further enhance the sensitivity and accuracy of cancer diagnostics. CRISPR-Cas9 can be used to enrich target sequences from complex samples, improving the detection of rare mutations and providing comprehensive genetic profiles of tumors. This integration allows for a more detailed understanding of tumor biology and the identification of novel therapeutic targets (Wan et al., 2017).

NGS provides high throughput sequencing capabilities, enabling the simultaneous analysis of thousands of genes and mutations. By integrating CRISPR-Cas9 with NGS, researchers can achieve targeted enrichment of specific genomic regions, enhancing the sensitivity and depth of sequencing. This combined approach facilitates the detection of low-frequency mutations and provides a comprehensive view of the genetic landscape of tumors (Wan et al., 2017).

Development of point-of-care diagnostics

The advancement of portable and user-friendly CRISPR-based diagnostic devices has great potential for cancer detection at the point of care. These gadgets offer prompt and precise outcomes, hence increasing their availability to a wider spectrum of patients and healthcare professionals. Point-of-care diagnostics can expedite the identification of diseases and prompt response, especially in situations with limited resources (Broughton et al., 2020).

Point-of-care diagnostics are specifically designed to be user-friendly, necessitating very basic training and little equipment. These technologies have the capability to provide results in a short time frame, ranging from minutes to hours, enabling prompt clinical decision-making. The objective of developing CRISPR-based point-of-care diagnostics is to provide cost-effective and easily obtainable cancer screening tools, thereby enhancing early diagnosis and minimizing inequalities in cancer treatment (Broughton et al., 2020).

Ethical and regulatory considerations

Ethical and regulatory considerations in CRISPR-Cas9 diagnostics

The integration of CRISPR-Cas9 technology into clinical diagnostics presents significant ethical and regulatory challenges that must be addressed to ensure its safe and effective use. The ethical considerations revolve around patient consent, data privacy, potential off-target effects, and the broader implications of genetic information. Regulatory considerations focus on the standardization of protocols, validation of diagnostic tools, and the establishment of guidelines to ensure accuracy and reliability. This section explores these ethical and regulatory issues in detail.

ETHICAL CONSIDERATIONS

Patient consent and autonomy

The development of portable and user-friendly CRISPR-based diagnostic equipment holds significant promise for detecting cancer at the point of care. These devices provide immediate and accurate results, thereby expanding their accessibility to a broader range of patients and healthcare professionals. Point-of-care diagnostics can accelerate the detection of diseases and facilitate a quick response, particularly in settings with few resources (Broughton et al., 2020).

Point-of-care diagnostics are specifically engineered to be easily used by users, requiring just minimal training and a small amount of equipment. These technologies have the capacity to deliver results within a relatively little time period, varying from minutes to hours, facilitating timely clinical decision-making. The aim of creating CRISPR-based point-of-care diagnostics is to provide affordable and readily accessible cancer screening technologies, which would improve early detection and reduce disparities in cancer treatment (Broughton et al., 2020).

Data privacy and security

The use of CRISPR-Cas9 in diagnostics involves the collection and analysis of sensitive genetic information, which raises significant privacy concerns. Protecting the confidentiality of genetic data is essential to prevent misuse or unauthorized access. Robust data security measures must be implemented to safeguard patient information, and clear policies must be established regarding data sharing and storage. Patients should be informed about how their genetic data will be used, who will have access to it, and how it will be protected (Ayday & Clayton, 2018).

Potential for discrimination

CRISPR-Cas9 diagnostics providing comprehensive genetic data may result in genetic discrimination by employers, insurers, or other groups. Regulations and legal structures, such as the Genetic Information Nondiscrimination Act (GINA) in the United States, are essential for safeguarding individuals against discrimination rooted in their genetic information (Rothstein, 2008). It is of utmost importance to uphold public trust in genetic technologies by ensuring responsible and ethical usage of genetic data.

Off-target effects and clinical implications

Although CRISPR-Cas9 is known for its accuracy, there is a possibility of off-target consequences, which involve unexpected modifications to genomic areas. The occurrence of these off-target effects might result in unforeseen outcomes, such as interference with crucial genes or the emergence of novel mutations. Off-target effects in diagnostics can lead to inaccurate test results, manifesting as either false positives or false negatives, so compromising the test's accuracy. In order to limit these dangers and assure reliable diagnostic conclusions, it is imperative to incorporate thorough validation and quality control methods into ethical considerations (Hsu et al., 2014).

Implications of incidental findings

CRISPR-Cas9 diagnostics can uncover unexpected findings, such as genetic predispositions to unrelated illnesses, alongside the original diagnostic inquiry. Ethically managing such findings entails alerting patients about the potential for incidental discoveries and establishing rules for handling and communicating these data. Genetic counseling is crucial in assisting patients in comprehending and managing the consequences of unexpected discoveries (Wolf et al., 2008).

REGULATORY CONSIDERATIONS

Standardization and validation

To ensure the accuracy and reliability of CRISPR-Cas9 diagnostics, standardized protocols and rigorous validation procedures must be established. This includes developing guidelines for gRNA design, CRISPR-Cas9 delivery methods, and the interpretation of diagnostic results. Regulatory bodies, such as the US Food and Drug Administration (FDA) and the European Medicines Agency (EMA), play a crucial role in setting these standards and overseeing the approval of CRISPR-based diagnostic tools (Shendure et al., 2017).

Quality control and assurance

Quality control measures are essential to maintain the integrity and consistency of CRISPR-Cas9 diagnostics. This involves implementing standardized procedures for sample preparation, CRISPR-Cas9 system assembly, and data analysis. Regular proficiency testing and inter-laboratory comparisons can help ensure that diagnostic tests meet established quality standards. Continuous monitoring and improvement of diagnostic processes are necessary to maintain high levels of accuracy and reliability (Berg et al., 2011).

Clinical implementation and training

Comprehensive training for healthcare workers is necessary for the successful incorporation of CRISPR-Cas9 diagnostics into clinical practice. This involves instructing clinicians, laboratory staff, and genetic counselors about the fundamental concepts of CRISPR technology, its uses, and its constraints. Implementing standardized training programs and certification processes can guarantee that healthcare personnel possess the necessary skills and knowledge to utilize CRISPR-based diagnostics with proficiency and ethical considerations (Shendure et al., 2017).

CONCLUSION

CRISPR-based diagnostic technologies have become effective tools for identifying and analyzing genetic alterations, infections, and biomarkers with exceptional accuracy and effectiveness. The CRISPR-Cas systems, which include Cas9, Cas12, and Cas13, have enabled progress in genome editing, illness diagnostics, and personalized treatment because of their adaptability. These technologies have shown considerable promise in clinical applications, including non-invasive prenatal testing, infectious illness detection, and cancer diagnostics. They provide fast, precise, and cost-efficient solutions. Nevertheless, the incorporation of CRISPR-based diagnostics into clinical practice gives rise to ethical and regulatory obstacles, such as guaranteeing informed permission, safeguarding data privacy, reducing off-target effects, and creating uniform methods and validation procedures. The ongoing advancements in CRISPR technology are anticipated to have a significant impact on healthcare, offering new prospects for early illness identification, individualized treatment strategies, and enhanced patient outcomes. The continuous progress highlights the significant impact that CRISPR-based diagnostics can have on transforming molecular diagnostics and developing precision medicine.

REFERENCES

Amado, R. G., Wolf, M., Peeters, M., Van Cutsem, E., Siena, S., Freeman, D. J., … Patterson, S. D. (2008). Wild-type KRAS is required for panitumumab efficacy in patients with metastatic colorectal cancer. Journal of Clinical Oncology, 26(10), 1626–1634.

Ayday, E., & Clayton, E. W. (2018). Privacy issues in genetic testing and personalized medicine. American Journal of Preventive Medicine, 54(5), 742–750.

Benn, P., Borrell, A., Chiu, R., Cuckle, H., Dugoff, L., Faas, B., … Wright, D. (2015). Position statement from the Chromosomal Abnormality Screening Committee on behalf of the Board of the International Society for Prenatal Diagnosis. Prenatal Diagnosis, 35(8), 725–734.

Berg, J. S., Khoury, M. J., & Evans, J. P. (2011). Deploying whole genome sequencing in clinical practice and public health: Meeting the challenge one bin at a time. Genetics in Medicine, 13(6), 499–504.

Broughton, J. P., Deng, X., Yu, G., Fasching, C. L., Servellita, V., Singh, J., ... Chiu, C. Y. (2020). CRISPR-Cas12-based detection of SARS-CoV-2. Nature Biotechnology, 38(7), 870–874.

Chen, J. S., Ma, E., Harrington, L. B., Da Costa, M., Tian, X., Palefsky, J. M., Doudna, J. A. (2018). CRISPR-Cas12a target binding unleashes indiscriminate single-stranded DNase activity. Science, 360(6387), 436–439.

Cong, L., Ran, F. A., Cox, D., Lin, S., Barretto, R., Habib, N., ... Zhang, F. (2013). Multiplex genome engineering using CRISPR/Cas systems. Science, 339(6121), 819–823.

Dever, D. P., Bak, R. O., Reinisch, A., Camarena, J., Washington, G., Nicolas, C. S., ... Porteus, M. H. (2016). CRISPR/Cas9 β-globin gene targeting in human haematopoietic stem cells. Nature, 539(7629), 384–389.

Doudna, J. A., & Charpentier, E. (2014). The new frontier of genome engineering with CRISPR-Cas9. Science, 346(6213), 1258096.

Freije, C. A., Myhrvold, C., Boehm, C. K., Lin, A. E., Welch, N. L., Carter, A., ... Sabeti, P. C. (2019). Programmable inhibition and detection of RNA viruses using Cas13. Molecular Cell, 76(5), 826–837.

Gasiunas G, Barrangou R, Horvath P, Siksnys V. (2012). Cas9-crRNA ribonucleoprotein complex mediates specific DNA cleavage for adaptive immunity in bacteria. Proceedings of the National Academy of Sciences of the United States of America. 109(39), E2579–E2586. https://doi.org/10.1073/pnas.1208507109

Gootenberg, J. S., Abudayyeh, O. O., Lee, J. W., Essletzbichler, P., Dy, A. J., Joung, J., ... Zhang, F. (2017). Nucleic acid detection with CRISPR-Cas13a/C2c2. Science, 356(6336), 438–442.

Hille, F., & Charpentier, E. (2016). CRISPR-Cas: Biology, mechanisms and relevance. Philosophical Transactions of the Royal Society B: Biological Sciences, 371(1707), 20150496.

Hsu, P. D., Lander, E. S., & Zhang, F. (2014). Development and applications of CRISPR-Cas9 for genome engineering. Cell, 157(6), 1262–1278.

Jinek, M., Chylinski, K., Fonfara, I., Hauer, M., Doudna, J. A., & Charpentier, E. (2012). A programmable dual-RNA–guided DNA endonuclease in adaptive bacterial immunity. Science, 337(6096), 816–821.

Komor, A., Kim, Y., Packer, M. et al. (2016). Programmable editing of a target base in genomic DNA without double-stranded DNA cleavage. Nature 533, 420–424. https://doi.org/10.1038/nature17946

Koo, B., Kim, M., Kim, G., Lee, J., Kim, S., Shin, S., & Park, C. (2021). Electrochemical biosensors for monitoring and point-of-care testing of infectious diseases. Biosensors and Bioelectronics, 181, 113053.

Kosicki, M., Tomberg, K., & Bradley, A. (2018). Repair of double-strand breaks induced by CRISPR-Cas9 leads to large deletions and complex rearrangements. Nature Biotechnology, 36(8), 765–771.

Levine, A. J., Oren, M., & Brash, D. E. (2019). The evolving landscape of p53 research: From basic biology to clinical implications. Cell, 178(2), 214–229.

Liu, F. X., Cui, J. Q., Wu, Z., & Yao, S. (2023). Recent progress in nucleic acid detection with CRISPR. Lab on a Chip, 23, 1467–1492.

Long, C., Amoasii, L., Mireault, A. A., McAnally, J. R., Li, H., Sanchez-Ortiz, E., ... Olson, E. N. (2018). Postnatal genome editing partially restores dystrophin expression in a mouse model of muscular dystrophy. Science, 351(6271), 400–403.

Lord, C. J., & Ashworth, A. (2016). BRCAness revisited. Nature Reviews Cancer, 16(2), 110–120.

Makarova, K. S., Wolf, Y. I., Alkhnbashi, O. S., Costa, F., Shah, S. A., Saunders, S. J., ... Koonin, E. V. (2020). An updated evolutionary classification of CRISPR–Cas systems. Nature Reviews Microbiology, 18(2), 67–83.

Ning, L., Li, Z., Tian, H., Zhang, S., Wu, W., & Yu, G. (2021). Multiplex CRISPR detection for diagnostics. Biotechnology Advances, 49, 107743.

O'Leary, B., Cutts, R. J., Liu, Y., Hrebien, S., Huang, X., Fenwick, K., ... Turner, N. C. (2018). The genetic landscape and clonal evolution of breast cancer resistance to palbociclib plus fulvestrant in PALOMA-3. Nature Communications, 9(1), 1–11.

Rothstein, M. A. (2008). Genetic information nondiscrimination act of 2008. PubMed Central, 69(2), 111–148.

Shendure, J., Findlay, G. M., & Snyder, M. W. (2017). Genomic medicine–progress, pitfalls, and promise. Cell, 168(6), 133–145.

Shmakov, S., Abudayyeh, O. O., Makarova, K. S., Wolf, Y. I., Gootenberg, J. S., Semenova, E., ... Koonin, E. V. (2017). Discovery and functional characterization of diverse Class 2 CRISPR-Cas systems. Molecular Cell, 60(3), 385–397.

Wang, Z., Liu, Y., Zhou, F. et al. (2023). The application of CRISPR-Cas in disease diagnosis and treatment. Science China Chemistry, 66, 2734–2742. https://doi.org/10.1007/s11426-023-1765-0
Wan, J. C., Massie, C., Garcia-Corbacho, J., Mouliere, F., Brenton, J. D., Caldas, C., ... Rosenfeld, N. (2017). Liquid biopsies come of age: Towards implementation of circulating tumor DNA. Nature Reviews Cancer, 17(4), 223–238.
Wolf, S. M., Annas, G. J., & Elias, S. (2008). Patient autonomy and incidental findings in clinical genomics. Science, 322(5903), 748–749.
Wright, A. V., Nuñez, J. K., & Doudna, J. A. (2016). Biology and applications of CRISPR systems: Harnessing nature's toolbox for genome engineering. Cell, 164(1–2), 29–44
Zhang, X., Sun, Y., Hsu, T. M., Chen, L., Wu, Q., Lam, H., ... Zhang, C. (2019). Tumor mutation burden is associated with response to immune checkpoint inhibitors in metastatic melanoma. Nature Medicine, 25(4), 512–523.

Chapter 5

Next-generation sequencing in molecular diagnostics

Qamre Alam[*] *and Kadhem Juma Alkhenaizi*
Molecular Genomics and Precision Medicine, ExpressMed Diagnostics and Research, Zinj, Kingdom of Bahrain

Misbahuddin M. Rafeeq
Department of Pharmacology, Faculty of Medicine, Rabigh, King Abdulaziz University, Jeddah, Saudi Arabia

Ali Bin Thani
Department of Biology, College of Science, University of Bahrain, Sakhir, Kingdom of Bahrain

Next-generation sequencing (NGS) is a major advancement in genomics that allows for the thorough examination of DNA and RNA. This technique has fundamentally transformed the approach to studying genetic material, offering an unparalleled understanding of the genetic and molecular foundations of health and illness. NGS technology is distinguished by its ability to process a large number of DNA fragments in a short amount of time, enabling the sequencing of millions of fragments simultaneously. This provides a comprehensive and extensive understanding of genomes. This capability has significantly advanced our understanding of complex genetic landscapes, facilitating breakthroughs in personalized medicine, genetic research, and molecular diagnostics.

The advent of NGS technology has effectively addressed several limitations inherent in previous sequencing technologies, notably Sanger sequencing. The capacity to rapidly and economically sequence substantial quantities of genetic material has created novel opportunities for research and clinical applications. Scientists are now able to perform tests without any preconceived hypotheses, where they can discover new genetic variations and biomarkers without knowing the particular target sequences in advance (Mardis, 2008). This has had a significant effect in fields such as cancer genomics, uncommon genetic disorders, and infectious disease diagnostics, where having complete genetic information is essential for understanding how diseases work and creating specific treatments.

The emergence of NGS is based on the development of sequencing technology. The initial technique, known as Sanger sequencing, was a pioneering technology created in the 1970s. However, it had notable drawbacks in terms of its capacity and expense (Heather & Chain, 2016). Sanger sequencing utilizes chain-termination techniques to sequence DNA fragments individually, resulting in a laborious and time-consuming process, especially when dealing with big genomes. Although Sanger sequencing had its limits, it was crucial in early genomic research, such as the Human Genome Project. This project, completed in 2003, yielded the first comprehensive map of the human genome (International Human Genome Sequencing Consortium, 2004).

[*] Corresponding author.

 DOI: 10.1201/9781003489177-5

The conclusion of the Human Genome Project emphasized the necessity for improved and expandable sequencing technology in order to properly exploit the capabilities of genetic data. As a result, NGS was developed in the early 2000s. NGS is a revolutionary technology that can do sequencing in a highly efficient manner by conducting many sequencing reactions simultaneously. This results in a significant improvement in speed and cost reduction compared to the traditional Sanger sequencing method (Mardis, 2008). The high-throughput sequencing of millions of DNA fragments concurrently was made possible by advancements in chemistry, engineering, and bioinformatics.

The commercialization of NGS technology was initiated by 454 Life Sciences in 2005, through the introduction of their pyrosequencing-based platform (Margulies et al., 2005). Subsequently, additional platforms emerged, such as Illumina's sequencing by synthesis method and Applied Biosystems' SOLiD sequencing. Each of these platforms employed distinct approaches for DNA sequencing, hence fostering innovation and competition in the industry. The tremendous progress in NGS technology in the last 20 years has greatly decreased the expense of sequencing, enabling its use in various scientific and medical areas (Mardis, 2013).

THE EVOLUTION AND IMPACT OF NGS

The influence of NGS on the field of genomics is of utmost significance. NGS has transformed our comprehension of genetic variation and disease by facilitating the fast and economical sequencing of whole genomes. NGS has had a profound effect on personalized medicine by enabling the customization of treatments according to an individual's genetic composition. In the field of cancer genomics, NGS is employed to detect genetic changes in cancers that can be effectively treated with certain medicines. This approach enhances therapy results and minimizes adverse effects (Koboldt et al., 2013).

NGS, among its medical implications, has also played a crucial role in enhancing our comprehension of fundamental biological processes. NGS has enhanced scientific study in fields including evolutionary biology, microbiology, and ecology by offering comprehensive knowledge about the genomic composition of species. Metagenomic sequencing has uncovered the extensive range of microbial communities in different settings, such as the human gut and the deep ocean. This has offered fresh perspectives on their functions in relation to health and disease (Qin et al., 2010).

In addition, NGS has played a crucial role in promoting the progress of precision medicine projects through the analysis of genetic variations in different populations. This has led to the discovery of genetic markers that are linked to diseases. As a result, predictive models and tailored interventions can be developed to address these disorders (Ashley, 2016). NGS has had a significant impact on the field of oncology. It is utilized to detect driver mutations in tumors, which helps in choosing targeted medicines and making treatment decisions (Mardis, 2012).

NGS WORKFLOW

The NGS workflow comprises multiple key steps, each important for producing sequence data of high quality. The process involves three main stages: sample preparation, sequencing, and data analysis (**Figure 5.1**). Every step of the process must be meticulously executed to ensure the accuracy and reliability of the sequencing results.

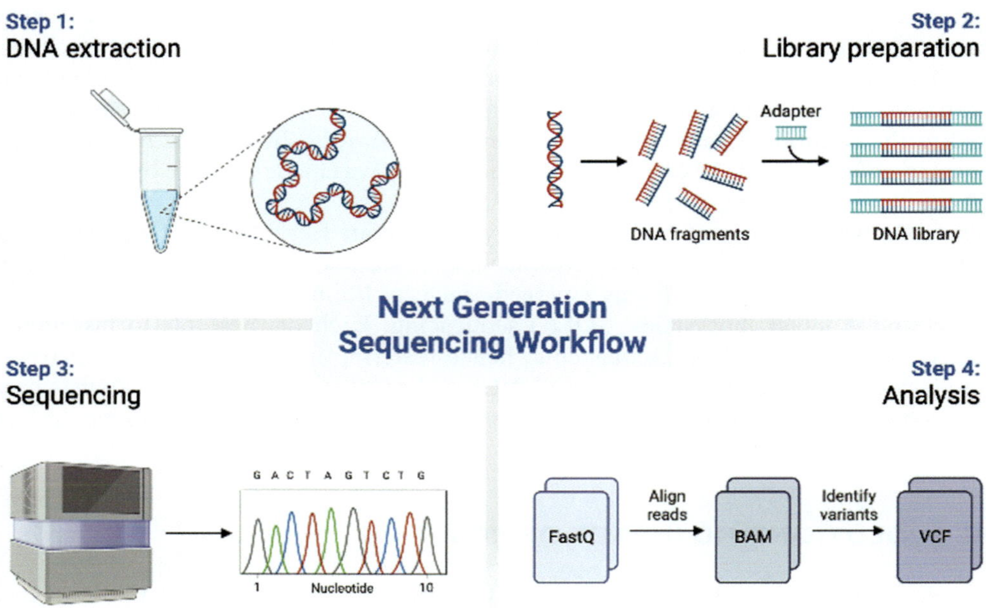

Figure 5.1 Diagram showing the steps in the Next Generation Sequencing (NGS) process.

Sample preparation

The initial stage of NGS includes the preparation of the DNA or RNA sample to be sequenced. This technique is essential since it establishes the fundamental basis for the entire sequencing workflow. The main steps involved in sample preparation include fragmentation, adapter ligation, and amplification.

Fragmentation: The DNA or RNA sample is broken down into smaller fragments to facilitate sequencing. This can be achieved through mechanical methods such as sonication or enzymatic treatments. Fragmentation is essential as it ensures that the sequencer can process the genetic material in smaller, uniform segments, which is necessary for accurate and efficient sequencing (Mardis, 2013).

Adapter ligation: After fragmentation, short DNA sequences known as adapters are ligated to the ends of the DNA fragments. These adapters have multiple functions: they establish a predetermined sequence for primers to bind to during amplification, facilitate the attachment of fragments to the sequencing platform, and include barcodes for sample identification, enabling the simultaneous sequencing of multiple samples in a single run (Bentley et al., 2008).

Amplification: The subsequent stage involves amplifying the DNA fragments via polymerase chain reaction (PCR). Amplification increases the amount of DNA, ensuring an adequate quantity of material for sequencing. PCR amplifies exclusively the fragments that have successfully linked adapters, hence increasing the concentration of sequences that are prepared for sequencing (Mardis, 2013). This step is crucial for amplifying the signal during sequencing, particularly when working with samples that have low input.

Sequencing

After the library has been created, the DNA fragments are placed onto the sequencing platform. There are multiple platforms that offer different approaches and technologies (**Table 5.1**). The selection of the platform is contingent upon the particular needs of the study, including the preferred read length, throughput, and accuracy.

Illumina sequencing: Illumina systems employ a technique known as sequencing by synthesis. Fluorescently tagged nucleotides are added to the DNA strand while it is being synthesized. With the addition of each nucleotide, a luminous signal is produced and identified by the sequencer. The DNA sequence is determined by the sequence of emitted fluorescence signals. This approach is characterized by its exceptional accuracy and ability to process large amounts of data quickly, which has led to its widespread adoption as one of the most often used NGS methods (Bentley et al., 2008).

Ion Torrent sequencing: Ion Torrent technology has a distinct methodology that relies on the identification of hydrogen ions generated during the integration of nucleotides. As nucleotides are incorporated into the DNA strand, the process results in the release of hydrogen ions, which subsequently leads to a change in the pH. The sequencer detects this alteration and determines the sequence according to the pH variations. Ion Torrent sequencing is renowned for its rapidity and comparatively affordable price, rendering it appropriate for targeted sequencing purposes (Rothberg et al., 2011).

Other platforms: Additional NGS technologies include Pacific Biosciences (PacBio), which provides the ability to sequence large reads, and Oxford Nanopore Technologies, which enables real-time sequencing and portability. Every platform possesses distinct advantages and is selected according to the specific requirements of the sequencing project (Goodwin, McPherson, & McCombie, 2016).

Data analysis

The NGS platforms produce extensive raw data that necessitates the use of advanced bioinformatics tools for analysis. The method of data processing is intricate and encompasses multiple important steps to ensure precise interpretation of the sequencing outcomes.

Table 5.1 Comparison of NGS platforms

NGS platform	*Principle*	*Read length*	*Throughput*	*Typical applications*
Illumina	Sequencing by synthesis	75–300 bp (paired-end)	Up to 6 Tb per run	Whole-genome sequencing, exome sequencing, RNA-seq
Ion Torrent	Sequencing by detection of pH changes	200–400 bp	Up to 15 Gb per run	Targeted sequencing, small genome sequencing
Pacific Biosciences (PacBio)	Single-molecule real-time sequencing	>10,000 bp (average)	Up to 100 Gb per run	Long-read sequencing, structural variant analysis
Oxford Nanopore	Nanopore sequencing	Up to >100,000 bp	Real-time sequencing, portable devices	Real-time sequencing, portable sequencing, field applications

Base calling: The first stage of data processing is base calling, which is the process of converting the raw signals (fluorescent signals or pH changes) into nucleotide sequences. The sequencer detects physical signals and turns them into digital nucleotide data, which may then be evaluated (Mardis, 2010).

Alignment and assembly: After the sequences are generated, they must be aligned to a reference genome or assembled de novo in the absence of a reference genome. Alignment is the process of mapping the short reads to a well-known reference genome, hence facilitating the identification of the precise location of each read within the genome. De novo assembly is the process of assembling small reads to rebuild the original sequence without using a reference. Identifying the structure and variations within the sequenced genome is a critical step (Mardis, 2010).

Variant calling: Following the process of alignment and assembly, the subsequent stage involves variant calling. This process entails the detection of disparities between the sequenced genome and the reference genome, including single-nucleotide polymorphisms, insertions, deletions (indels), and structural variants. Variant calling necessitates the use of advanced algorithms and bioinformatics tools to precisely identify and characterize genomic changes (Mardis, 2013).

Data interpretation: Data interpretation is the last stage in the NGS workflow. This entails examining the identified genetic variations to understand their biological importance. In clinical settings, this could involve establishing connections between genetic variations and disease characteristics, pinpointing possible targets for treatment, or evaluating the effects of mutations on protein functionality. To achieve a thorough understanding of the genetic landscape, it is often necessary to combine NGS data with other biological and clinical data. This integration is crucial for effective data interpretation (Koboldt et al., 2013).

NGS METHODOLOGIES

Various NGS approaches are employed based on the specific requirements of the research or diagnostic purposes. These approaches differ in how they sequence genetic material, offering versatility in terms of the level of detail, depth, and comprehensiveness of genetic study. There are two main approaches used in this context: amplicon sequencing and shotgun metagenomic sequencing.

Amplicon sequencing

Amplicon sequencing is a method that entails amplifying targeted sections of the genome, typically using PCR, prior to sequencing. This approach is especially valuable for focused sequencing applications that prioritize certain genomic regions of interest.

Targeted amplification: Amplicon sequencing is a technique that specifically amplifies and analyzes targeted sections of the genome. Typically, this is accomplished by creating primers that surround the specific sections that need to be sequenced. One frequent use is the examination of microbial communities through the process of 16S rRNA gene sequencing. The 16S rRNA gene is extremely conserved in bacteria, but it also contains sections that vary greatly and can be used to identify individual species. This makes it well-suited for bacterial identification and classification (Caporaso et al., 2012).

Amplicon sequencing offers a significant benefit in terms of cost-effectiveness. By selectively focusing on particular regions, the quantity of sequencing needed is reduced, resulting in a decrease in the overall expense. Moreover, this approach offers extensive coverage of the specific areas of interest, hence improving the precision and dependability of the acquired sequence data. Extensive coverage is especially advantageous when researching species with low numbers or uncommon genetic variations (Sinclair et al., 2015).

Although amplicon sequencing offers benefits, it also has limitations. The detection is restricted to the sequences that are surrounded by the primer binding sites, hence any alterations occurring outside of these regions cannot be identified. Moreover, the dependence on PCR amplification might lead to biases, as certain sequences may be selectively amplified, thus influencing the outcomes. The design of primers is crucial since improperly designed primers might lead to non-specific amplification or the inability to amplify the desired region (Poretsky et al., 2014).

Shotgun metagenomic sequencing

Shotgun metagenomic sequencing differs from amplicon sequencing in that it involves the random fragmentation of the whole genome, followed by sequencing of the fragments. This methodology offers a more extensive insight into the genetic material, enabling the detection and identification of all genes that are present in a given sample.

Shotgun metagenomic sequencing utilizes random sequencing of the complete genome, without relying on focused amplification. This facilitates a more thorough examination of the genetic material, hence permitting the identification of all genes that are present in a given sample. This technology is highly advantageous for investigating intricate microbial communities, since it has the capability to detect a diverse array of microorganisms, even some that are challenging to cultivate in the laboratory (Moran et al., 2013).

An important benefit of shotgun sequencing is its capacity to uncover previously unknown genes and metabolic pathways. Through the process of sequencing the complete genetic material, scientists are able to discover genes that were previously unknown and obtain a deeper understanding of the functional capabilities of microbial communities. These findings have significant significance for comprehending microbial ecology and the functions of different species in diverse habitats (Qin et al., 2010).

Shotgun metagenomic sequencing is also efficient in identifying infrequent genetic variations. By sequencing the entire genome, even sequences that are present in tiny amounts can be detected, allowing for a more comprehensive understanding of the genetic variation present in a sample. Studying genetic variation and evolution requires special attention, especially when it comes to rare variants that can have profound biological consequences (Tyson et al., 2004).

Although shotgun sequencing is extensive, it also presents obstacles. The enormous volume of data collected necessitates significant computational resources for analysis and storage. An understanding of advanced bioinformatics technologies is crucial for effectively managing and accurately interpreting data. Moreover, the inherent randomness of sequencing implies that the distribution of coverage may be uneven, resulting in certain sections being read with greater depth than others. Complex or repeated areas can pose challenges in the assembly and annotation of genomes, as highlighted by Venter et al. (2004).

ADVANCEMENTS IN NGS

The field of NGS technology has shown substantial progress since its inception, resulting in improved precision, speed, and cost-effectiveness. These developments have overcome multiple constraints of previous sequencing technology, enabling more thorough and accurate genetic analysis.

LONG-READ SEQUENCING TECHNOLOGIES

A significant breakthrough in NGS is the emergence of long-read sequencing technologies (**Table 5.1**). In contrast to short-read sequencing, which produces reads consisting of only a few hundred base pairs, long-read sequencing creates reads that can span thousands of base pairs. This capability is especially useful for identifying repetitive sections and intricate structural differences within genomes, which can be difficult to examine with short-read methods.

Pacific Biosciences (PacBio): PacBio's Single Molecule Real-Time (SMRT) sequencing technique allows for the production of lengthy reads with remarkable precision. This method enables the sequencing of DNA in real time, facilitating the identification of changes in the DNA bases and the analysis of intricate genomic regions. PacBio's extended reads play a crucial role in enhancing genome assemblies, detecting structural variations, and examining epigenetic alterations (Eid et al., 2009; English et al., 2012).

Oxford Nanopore Technologies: Oxford Nanopore's sequencing technology uses nanopores to directly read DNA sequences as they traverse through a nanopore channel. This method enables the generation of extremely long readings, occasionally exceeding a length of 100 kilobases. Nanopore sequencing is exceptionally portable, allowing for immediate genomic analysis in many environments, ranging from clinical diagnostics to field research (Jain et al., 2016; Quick et al., 2016).

The utilization of long-read sequencing technologies has greatly improved our capacity to construct genomes from scratch, precisely identify structural differences, and investigate intricate regions such as centromeres and telomeres. These breakthroughs have wide-ranging implications for studies in genetics, evolutionary biology, and medicine.

Single-Cell Sequencing

Another significant advancement in NGS is single-cell sequencing, which allows for the analysis of the genomic and transcriptomic profiles of individual cells (**Figure 5.2**). This technique has revolutionized fields such as cancer research, developmental biology, and

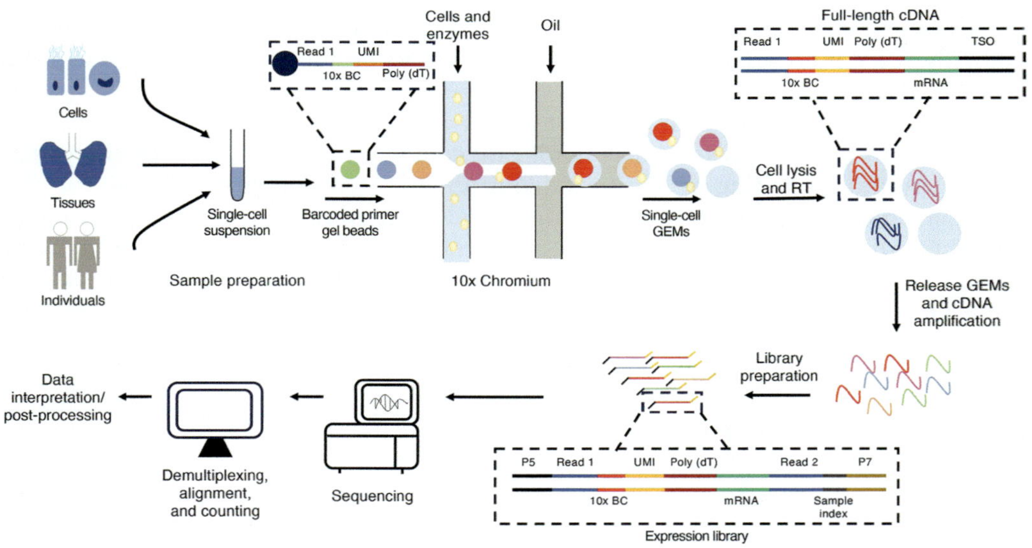

Figure 5.2 Illustration depicting the workflow for single-cell sequencing.

immunology by providing insights into cellular heterogeneity and the clonal evolution of tumors.

Genomic profiling: Single-cell sequencing enables the identification of genetic variations and mutations at the single-cell level. This is particularly valuable in cancer research, where it can reveal the clonal evolution of tumors and the presence of rare cell populations that drive disease progression and treatment resistance (Navin et al., 2011; Patel et al., 2014).

Transcriptomic profiling: Single-cell RNA sequencing (scRNA-seq) allows for the comprehensive analysis of gene expression profiles in individual cells. This technique has provided new insights into cellular differentiation, tissue heterogeneity, and the dynamic responses of cells to various stimuli. It has been instrumental in mapping the cellular composition of complex tissues and identifying novel cell types (Tang et al., 2009; Macosko et al., 2015).

The ability to analyze individual cells at high resolution has uncovered previously unrecognized aspects of cellular function and heterogeneity, driving new discoveries in biology and medicine.

HIGH-THROUGHPUT AND COST EFFICIENCY

NGS technologies have undergone substantial advancements in both the speed at which they can process genetic data and their ability to do it at a lower cost. The cost per base of sequencing has been reduced, and the amount of data collected each run has increased due to advancements in sequencing chemistry, apparatus, and bioinformatics.

High-throughput sequencing: Contemporary NGS platforms, including Illumina's NovaSeq, have the capability to produce terabases of data in a single operation. The ability to process a large number of samples quickly enables the sequencing of huge groups of individuals and broad investigations of the genome. This makes it easier to carry out projects involving large populations and studies of the link between diseases and genetic factors (Bentley et al., 2008; Goodwin, McPherson, & McCombie, 2016).

Cost reduction: The cost of sequencing has dropped dramatically since the introduction of NGS. The development of more efficient sequencing chemistries and automation has contributed to this reduction, making genomic research more accessible. This affordability has enabled a broader range of applications, from routine clinical diagnostics to large-scale research initiatives like the 100,000 Genomes Project (Mardis, 2013).

INTEGRATION OF BIOINFORMATICS

The use of sophisticated bioinformatics tools has played a vital role in handling and analyzing the vast volumes of data produced by NGS. Advancements in computing algorithms, data storage, and analytical frameworks have improved the precision and effectiveness of genetic investigations.

Advanced bioinformatics pipelines have been created to manage processes such as read alignment, variant calling, and genome assembly. The utilization of these methods is crucial for converting unprocessed sequencing data into significant biological insight (Li & Durbin, 2009; DePristo et al., 2011).

Machine learning and artificial intelligence (AI) have enhanced the analysis of NGS data in genomics. Machine learning algorithms have the ability to detect patterns and relationships in intricate datasets, which can assist in uncovering new biomarkers and predicting the outcomes of diseases (Libbrecht & Noble, 2015).

The progress made in the field of bioinformatics has played a crucial role in the achievement of NGS, allowing for the retrieval of meaningful data from extensive and intricate genomic databases.

APPLICATIONS OF NGS

NGS has extensive applications in various fields like health, biology, and environmental science. This technology has significantly transformed various parts of research and clinical practice by offering an extensive and precise understanding of genetic material (**Figure 5.3**). NGS offers distinct advantages to several fields by providing high-speed processing, precise results, and comprehensive data. This technology plays a crucial role in advancing our knowledge and control of intricate biological and environmental systems (**Table 5.2**).

Genetic disorders

NGS facilitates a thorough examination of the human genome, enabling the detection of genetic abnormalities linked to hereditary illnesses. A widely used application in this field is whole exome sequencing (WES), which focuses on the protein-coding portions of the genome. The exons, which make up only 1% of the genome, include around 85% of the known disease-related variations (Ng et al., 2010). WES is highly valuable in the diagnosis of uncommon genetic illnesses. It has the ability to detect both new and previously identified mutations that have a role in the disease phenotype. This has been extremely beneficial for people with undiscovered illnesses, offering valuable data that can inform clinical treatment and genetic counseling.

Diagnostic tool

WES is utilized for the diagnosis of many genetic illnesses, encompassing neurodevelopmental abnormalities, metabolic issues, and hereditary malignancies. Through the process of sequencing the exomes of individuals who are experiencing symptoms, researchers are able to identify specific mutations that could potentially be the cause of the observed clinical symptoms. The exact identification of causal mutations can result in accurate diagnoses and

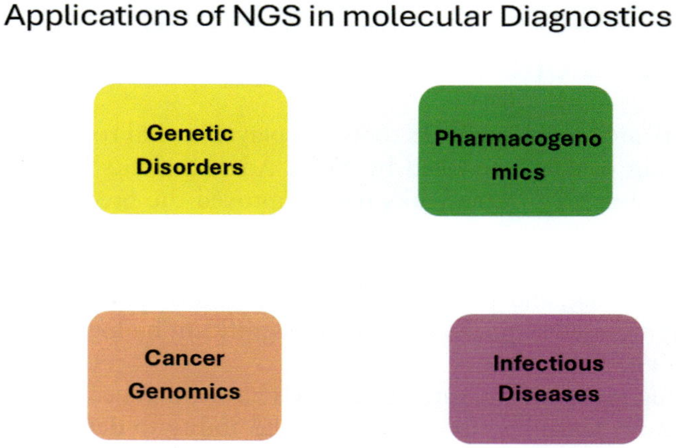

Figure 5.3 Flowchart showcasing various applications of Next Generation Sequencing in molecular diagnostics.

Table 5.2 Applications of NGS in clinical diagnostics

Application	*Methodology*	*Benefits*	*Examples*
Genetic disorders	Whole exome sequencing	Identification of disease-causing mutations	Diagnosing rare genetic disorders, identifying novel mutations
Cancer genomics	Targeted sequencing, Whole genome sequencing, liquid biopsies	Personalized treatment plans, monitoring tumor evolution	Identifying EGFR mutations in NSCLC, monitoring KRAS mutations in colorectal cancer
Infectious disease diagnostics	Metagenomic sequencing	Comprehensive pathogen identification	Detecting SARS-CoV-2 in respiratory samples, identifying bacterial infections in sepsis
Pharmacogenomics	Targeted sequencing of pharmacogenes	Personalized medication strategies	Identifying CYP2C19 variants affecting drug metabolism, guiding antidepressant selection
Prenatal testing	Non-invasive prenatal testing	Early detection of chromosomal abnormalities	Detecting trisomy 21 (Down syndrome), trisomy 18 (Edwards syndrome)
Carrier screening	Targeted sequencing	Risk assessment for inherited conditions	Screening for cystic fibrosis carrier status, identifying carriers of Tay-Sachs disease

individualized treatment strategies, which can greatly enhance patient outcomes (Boycott et al., 2013). For instance, WES has revealed mutations in genes that were previously unrelated to neurodevelopmental disorders in patients, thereby expanding the understanding of their genetic basis (Deciphering Developmental Disorders Study, 2015).

Research and discovery

NGS expedites the identification of novel disease genes. Researchers can find new genetic variations linked to specific ailments by comparing the exomes of affected individuals with those of healthy controls. As a result, we have been able to identify a large number of genes that cause diseases and this has greatly enhanced our knowledge of the genetic factors underlying many disorders (Bamshad et al., 2011). Studies have discovered novel genes associated with illnesses such as epilepsy, intellectual disability, and cardiomyopathies. These findings offer potential targets for therapeutic intervention (Ropers, 2010; Helbig et al., 2016).

Carrier screening and prenatal testing

NGS is also employed for carrier screening and prenatal testing. Carrier screening is a method used to identify individuals who possess one copy of a gene mutation. This mutation, when present in two copies, leads to the development of a genetic condition. Prospective parents must understand the significance of this knowledge when evaluating their likelihood of having a kid with a genetic disorder (Nazareth et al., 2016). Prenatal testing utilizing NGS has the capability to identify genetic anomalies in a developing fetus, allowing for prompt detection and treatment of genetic illnesses (Bianchi et al., 2014).

Pharmacogenomics

Pharmacogenomics, a major application of NGS in genetic illnesses, investigates the impact of genes on an individual's pharmacological response. NGS can be used to detect genetic variations that impact drug metabolism, effectiveness, and safety. This information can then be used to guide personalized medicine strategies, ensuring that patients receive the most suitable and secure pharmaceuticals depending on their genetic characteristics (Daly, 2010).

CANCER GENOMICS

In oncology, NGS is used to identify somatic mutations in tumors, guiding the selection of targeted therapies. The ability to perform liquid biopsies, where cell-free DNA (cfDNA) from tumors is sequenced from blood samples, has revolutionized cancer diagnostics and monitoring.

Targeted therapies: NGS allows for the identification of genetic mutations within tumors, enabling the development of targeted therapies that specifically address the genetic abnormalities driving cancer growth. This personalized approach improves treatment efficacy and reduces adverse effects compared to traditional chemotherapy (Koboldt et al., 2013). For example, NGS has identified mutations in the EGFR gene in non-small cell lung cancer, which can be targeted with tyrosine kinase inhibitors (TKIs), improving patient outcomes (Lynch et al., 2004).

Liquid biopsies: Liquid biopsies represent a non-invasive method to monitor cancer progression and treatment response. By analyzing cfDNA in the blood, NGS can detect genetic mutations and monitor changes over time, providing real-time insights into the tumor's genetic landscape. This is particularly valuable for tracking minimal residual disease and detecting early signs of relapse (Bettegowda et al., 2014). Liquid biopsies have been used to monitor the effectiveness of treatments in cancers such as colorectal, breast, and prostate cancer, providing a dynamic view of tumor evolution (Wan et al., 2017).

Molecular profiling: Comprehensive molecular profiling of tumors using NGS provides a detailed understanding of the genetic alterations present in cancer. This information can be used to classify tumors, predict prognosis, and identify potential therapeutic targets. For instance, The Cancer Genome Atlas (TCGA) project has utilized NGS to catalog genetic mutations in various cancer types, leading to new insights into cancer biology and treatment strategies (Weinstein et al., 2013).

Drug resistance: NGS also helps in understanding mechanisms of drug resistance in cancer. By sequencing tumors before and after treatment, researchers can identify genetic changes that confer resistance to therapies, allowing for the development of strategies to overcome resistance and improve patient outcomes (Diaz et al., 2012).

INFECTIOUS DISEASES

NGS is also utilized for the identification and analysis of infections in clinical samples. Metagenomic sequencing enables the detection of bacteria, viruses, fungi, and parasites in a single test, offering a full perspective of the microbial environment.

Identification of pathogens: Conventional approaches to pathogen identification frequently depend on culture techniques, which can be laborious and may fail to detect all pathogens. NGS, in contrast, has the ability to quickly analyze the complete genetic information in a sample, accurately detecting both known and new diseases with great sensitivity and specificity (Miller et al., 2019). NGS has been utilized to find new viruses in respiratory illnesses and to detect mixed infections that may not be detected by conventional approaches (Greninger et al., 2015).

Epidemiological surveillance: NGS is used in public health to track the spread of infectious diseases. By sequencing the genomes of pathogens, researchers can monitor transmission patterns, detect outbreaks, and understand the evolution of antimicrobial resistance (AMR). This information is critical for guiding public health interventions and developing strategies to control infectious diseases (Miller et al., 2019). For example, during the COVID-19 pandemic, NGS has been used extensively to track the spread of SARS-CoV-2, identify variants of concern, and inform public health responses (Candido et al., 2020).

Antimicrobial resistance: NGS plays a crucial role in studying AMR. By sequencing bacterial genomes, researchers can identify resistance genes and understand the mechanisms by which bacteria acquire and spread resistance. This knowledge is essential for developing new antibiotics and strategies to combat AMR (Dunne et al., 2017).

Microbial ecology: Beyond human health, NGS is used to study microbial communities in various environments, including soil, water, and food. Understanding the diversity and function of microbial communities is important for ecosystem management, agriculture, and food safety (Thompson et al., 2017).

CASE STUDIES OF NGS IN MOLECULAR DIAGNOSTICS

The integration of NGS into molecular diagnostics has significantly advanced the understanding and treatment of various diseases. Below are detailed case studies that exemplify the transformative impact of NGS in different medical contexts.

Diagnosing rare genetic disorders

One specific example involved a 7-year-old child who had significant cognitive impairment, delays in development, and frequent seizures. Despite using conventional diagnostic techniques such as karyotyping and single-gene testing, the origin of his condition remained unknown. As a result, the boy underwent WES to find the underlying cause. WES identified a new mutation in the SCN2A gene, which is responsible for encoding a crucial sodium channel protein necessary for proper neuronal activity. SCN2A mutations are linked to a range of neurological diseases, such as epileptic encephalopathies. The discovery of this information resulted in a conclusive identification of the condition and enabled the use of specific medications that block sodium channels. This treatment approach effectively decreased the occurrence and intensity of the boy's seizures (Ng et al., 2010; Bamshad et al., 2011). This instance highlights the efficacy of NGS in diagnosing uncommon genetic abnormalities that are difficult to identify using conventional diagnostic techniques.

Precision oncology

NGS has become an essential tool in the field of oncology, offering comprehensive genetic data that informs individualized treatment strategies. As an illustration, a 55-year-old female with advanced non-small cell lung cancer (NSCLC) initially exhibited partial remission with standard chemotherapy but experienced a recurrence of her cancer within a span of six months. NGS was used to analyze a specific set of genes associated with cancer. This analysis revealed the presence of a mutation in the EGFR gene that enhances the response to TKIs such as gefitinib and erlotinib. The patient's treatment was changed to gefitinib therapy based on the results of these studies. This led to a remarkable response, with a notable reduction in tumor size and better clinical symptoms (Lynch et al., 2004; Koboldt et al., 2013). This instance emphasizes the crucial significance of NGS in precision oncology, facilitating individualized treatment strategies that enhance patient outcomes.

Infectious disease diagnosis

NGS has also demonstrated its critical role in the diagnosis and management of infectious illnesses. A notable case involved a 45-year-old male patient who was taken to the hospital due to acute respiratory symptoms, such as high body temperature, persistent cough, and difficulty breathing. The preliminary screenings for prevalent respiratory infections, such as influenza and respiratory syncytial virus, yielded negative results. Metagenomic NGS analysis conducted on a sample obtained from a bronchoalveolar lavage revealed the presence of a previously unidentified coronavirus. This novel coronavirus was subsequently called SARS-CoV-2 and was determined to be the cause of the patient's symptoms. The early detection of COVID-19 was crucial in identifying the epidemic and allowed for prompt public health interventions, such as isolation and contact tracing, to mitigate the transmission of the virus (Greninger et al., 2015; Miller et al., 2019). This instance exemplifies the crucial role of NGS in promptly identifying and describing newly emerging infectious illnesses.

Monitoring cancer treatment with liquid biopsies

NGS is being employed for non-invasive monitoring of cancer treatment using liquid biopsies. A 60-year-old man with metastatic colorectal cancer originally exhibited a positive response to targeted therapy. However, he subsequently developed resistance, resulting in the development of the disease. The analysis of cell-free DNA (cfDNA) in the patient's blood using liquid biopsy NGS revealed the presence of a KRAS mutation, which is known to cause resistance to EGFR-targeted therapy. The absence of this mutation in the first tumor biopsy suggests clonal evolution and resistance to treatment. According to these discoveries, the patient's treatment plan was modified to address the newly identified genetic changes, resulting in better results (Bettegowda et al., 2014; Wan et al., 2017). This instance emphasizes the significance of liquid biopsy NGS in the ongoing assessment of cancer and providing guidance for treatment modifications.

CHALLENGES AND FUTURE DIRECTIONS

Despite its transformative potential, NGS faces several challenges that need to be addressed to maximize its benefits and integrate it seamlessly into clinical and research settings.

Challenges

A significant challenge related to NGS is the comprehension of the vast volume of data it produces. Every sequencing run has the capacity to generate terabytes of data, encompassing not just the main sequences but also a wide range of variations, structural alterations, and other genomic characteristics. The bioinformatics analyses necessary to handle and interpret these data sets are highly sophisticated. Precise variant calling, alignment, and annotation are crucial stages that necessitate advanced computational techniques and specialized skills (Schloss et al., 2011).

Ensuring the production of sequencing data of superior quality necessitates the effective management of several issues, including sequencing errors, biases in library formation, and contamination. PCR amplification may create errors or biases, especially in regions with a high GC content or repetitive sequences. Data integrity can be compromised by contaminants from environmental sources or previous experiments, which can result in false positives or incorrect findings (Li & Durbin, 2009; Schloss et al., 2011).

There is a significant need for strong bioinformatics infrastructure because of the huge amount of data generated by NGS. This encompasses state-of-the-art computing systems for efficient data processing, cutting-edge storage solutions for effective data management, and advanced bioinformatics tools for comprehensive data analysis. Building and sustaining this infrastructure can be financially impractical, especially for smaller research institutes or clinical laboratories. Furthermore, there is an ongoing demand for proficient bioinformaticians who possess the ability to proficiently create, enhance, and utilize these technologies (Mardis, 2010).

Standardizing NGS protocols is essential to ensure the reproducibility and accuracy of results across different laboratories. Variations in sample preparation, sequencing platforms, and data analysis methods can lead to significant discrepancies in results. Establishing standardized protocols and quality control measures is critical for achieving consistent and reliable outcomes. This includes standardizing the methods for library preparation, sequencing, and data processing, as well as developing universal benchmarks for performance evaluation (DePristo et al., 2011; Manolio et al., 2013).

Integrating NGS into clinical practice requires navigating a complex landscape of regulatory and ethical issues. Regulatory bodies need to establish guidelines for the clinical use of NGS, including validation and quality assurance of sequencing tests. Ethical considerations include ensuring patient consent, data privacy, and the potential for incidental findings, where unexpected genetic information with significant health implications is discovered. Addressing these issues requires collaboration between scientists, clinicians, ethicists, and regulatory authorities (Wolf et al., 2008; Manolio et al., 2013).

Implementing NGS in clinical settings involves additional challenges, such as integrating sequencing data into electronic health records and ensuring that clinicians have the necessary training to interpret and act on genomic information. The clinical utility of NGS results must be clearly demonstrated to justify their use in routine clinical practice. This includes validating the clinical relevance of identified variants and developing guidelines for their interpretation and management (Manolio et al., 2013; Shendure et al., 2017).

Future directions

Future advancements in bioinformatics will play a crucial role in overcoming the challenges of NGS data interpretation. This includes the development of more sophisticated algorithms for variant calling, alignment, and annotation, as well as machine learning approaches to predict the functional impact of genetic variants. Improvements in computational efficiency will enable faster and more accurate analysis of large-scale genomic data sets (Libbrecht & Noble, 2015).

Integrating NGS data with other "omics" data, such as transcriptomics, proteomics, and metabolomics, will provide a more comprehensive understanding of biological systems. This multi-omics approach can reveal complex interactions between different molecular layers, offering deeper insights into disease mechanisms and potential therapeutic targets. Advances in data integration and analysis tools will be essential for realizing the full potential of multi-omics research (Hasin et al., 2017).

The development of long-read sequencing technologies, such as those offered by Pacific Biosciences (PacBio) and Oxford Nanopore, promises to address some of the limitations of short-read sequencing. Long-read technologies can provide more accurate genome assemblies, particularly in regions with complex structural variations or repetitive sequences. These advancements will enhance the ability to detect and characterize genomic alterations, improving the diagnosis and treatment of genetic disorders (Goodwin et al., 2016).

The application of NGS in personalized medicine will continue to expand, with increasing use in diagnostics, prognostics, and treatment selection. Ongoing research will further elucidate the clinical relevance of genetic variants, leading to more precise and individualized treatment strategies. NGS will also play a crucial role in the development of new therapies, including gene editing and personalized vaccines (Ashley, 2016).

Future endeavors will prioritize the development of thorough ethical and regulatory frameworks for the clinical application of NGS. This encompasses the creation of protocols for obtaining patient consent, ensuring data privacy, and handling incidental results. In order to ensure the reliability and safety of NGS testing in clinical practice, regulatory agencies must establish guidelines for their validation and quality assurance (Wolf et al., 2008; Shendure et al., 2017).

International collaboration will be crucial for progressing NGS research and applications. The Global Alliance for Genomics and Health (GA4GH) seeks to facilitate the exchange of genetic data and promote the use of optimal methods and strategies on a global scale. It is crucial to make NGS technology more accessible, especially in low-resource settings, in order to ensure that the advantages of genomic medicine are realized globally (Manolio et al., 2013; The 1000 Genomes Project Consortium, 2015).

CONCLUSION

NGS has transformed genomics by allowing for efficient and thorough examination of DNA and RNA. This has greatly accelerated genetic research and molecular diagnostics. This technique has advanced beyond the constraints of Sanger sequencing to enable fast and cost-efficient sequencing abilities that offer a comprehensive understanding of the genetic basis of health and illness. The NGS methodology, encompassing sample preparation to data analysis, guarantees precise sequencing outcomes, while innovations like long-read and single-cell sequencing have expanded its utility in diverse domains, such as personalized medicine, cancer genomics, and infectious disease diagnostics. Although there are difficulties in interpreting and controlling the quality of data, ongoing progress and strong support in the field of bioinformatics are expected to enhance the integration of NGS into clinical practice. This will contribute to advancing our knowledge and treatment of genetic disorders, malignancies, and infectious diseases.

REFERENCES

Ashley, E. A. (2016). Towards precision medicine. Nature Reviews Genetics, 17(9), 507–522.

Bamshad, M. J., Ng, S. B., Bigham, A. W., Tabor, H. K., Emond, M. J., Nickerson, D. A., & Shendure, J. (2011). Exome sequencing as a tool for Mendelian disease gene discovery. Nature Reviews Genetics, 12(11), 745–755.

Bentley, D. R., Balasubramanian, S., Swerdlow, H. P., Smith, G. P., Milton, J., Brown, C. G., … Smith, A. J. (2008). Accurate whole human genome sequencing using reversible terminator chemistry. Nature, 456(7218), 53–59.

Bettegowda, C., Sausen, M., Leary, R. J., Kinde, I., Wang, Y., Agrawal, N., … Diaz, L. A. (2014). Detection of circulating tumor DNA in early- and late-stage human malignancies. Science Translational Medicine, 6(224), 224ra24.

Bianchi DW, Parker RL, Wentworth J, Madankumar R, Saffer C, Das AF, Craig JA, Chudova DI, Devers PL, Jones KW, Oliver K, Rava RP, Sehnert AJ; CARE Study Group. DNA sequencing versus standard prenatal aneuploidy screening. N Engl J Med. 2014;370(9), 799–808

Boycott, K. M., Vanstone, M. R., Bulman, D. E., & MacKenzie, A. E. (2013). Rare-disease genetics in the era of next-generation sequencing: Discovery to translation. Nature Reviews Genetics, 14(10), 681–691.

Candido, D. S., et al. (2020). Evolution and epidemic spread of SARS-CoV-2 in Brazil. Science, 369(6508), 1255–1260.

Caporaso, J. G., Lauber, C. L., Walters, W. A., Berg-Lyons, D., Lozupone, C. A., Turnbaugh, P. J., ... Knight, R. (2012). Global patterns of 16S rRNA diversity at a depth of millions of sequences per sample. Proceedings of the National Academy of Sciences, 108(Supplement 1), 4516–4522.

Daly, A. K. (2010). Pharmacogenetics and human genetic polymorphisms. Biochemical Journal, 429(3), 435–449.

Deciphering Developmental Disorders Study. (2015). Large-scale discovery of novel genetic causes of developmental disorders. Nature, 519(7542), 223–228.

DePristo, M. A., Banks, E., Poplin, R., Garimella, K. V., Maguire, J. R., Hartl, C., ... Daly, M. J. (2011). A framework for variation discovery and genotyping using next-generation DNA sequencing data. Nature Genetics, 43(5), 491–498.

Diaz, L. A., et al. (2012). The molecular evolution of acquired resistance to targeted EGFR blockade in colorectal cancers. Nature, 486(7404), 537–540.

Dunne, W. M., et al. (2017). Next-generation sequencing in clinical microbiology: Are we there yet? Clinical Microbiology Reviews, 30(1), 121–141.

Eid, J., Fehr, A., Gray, J., Luong, K., Lyle, J., Otto, G., ... Turner, S. (2009). Real-time DNA sequencing from single polymerase molecules. Science, 323(5910), 133–138.

English, A. C., Richards, S., Han, Y., Wang, M., Vee, V., Qu, J., ... Gibbs, R. A. (2012). Mind the gap: Upgrading genomes with Pacific Biosciences RS long-read sequencing technology. PloS One, 7(11), e47768.

Goodwin, S., McPherson, J. D., & McCombie, W. R. (2016). Coming of age: Ten years of next-generation sequencing technologies. Nature Reviews Genetics, 17(6), 333–351.

Greninger, A. L., et al. (2015). Metagenomic next-generation sequencing of plasma cell-free DNA from patients with febrile illness and respiratory infection. Journal of Clinical Microbiology, 53(7), 2162–2165.

Hasin, Y., Seldin, M., & Lusis, A. (2017). Multi-omics approaches to disease. Genome Biology, 18(1), 83.

Heather, J. M., & Chain, B. (2016). The sequence of sequencers: The history of sequencing DNA. Genomics, 107(1), 1–8.

Helbig, I., et al. (2016). Genetic determinants of epilepsy: A review. Neuropsychopharmacology, 41(1), 89–112.

International Human Genome Sequencing Consortium. (2004). Finishing the euchromatic sequence of the human genome. Nature, 431(7011), 931–945.

Jain, M., Olsen, H. E., Paten, B., & Akeson, M. (2016). The Oxford Nanopore MinION: Delivery of nanopore sequencing to the genomics community. Genome Biology, 17(1), 239.

Koboldt, D. C., Steinberg, K. M., Larson, D. E., Wilson, R. K., & Mardis, E. R. (2013). The next-generation sequencing revolution and its impact on genomics. Cell, 155(1), 27–38.

Li, H., & Durbin, R. (2009). Fast and accurate short read alignment with Burrows–Wheeler transform. Bioinformatics, 25(14), 1754–1760.

Libbrecht, M. W., & Noble, W. S. (2015). Machine learning applications in genetics and genomics. Nature Reviews Genetics, 16(6), 321–332.

Lynch, T. J., et al. (2004). Activating mutations in the epidermal growth factor receptor underlying responsiveness of non-small-cell lung cancer to gefitinib. New England Journal of Medicine, 350(21), 2129–2139.

Manolio, T. A., Chisholm, R. L., Ozenberger, B., Roden, D. M., Williams, M. S., Wilson, R., ... & Ginsburg, G. S. (2013). Implementing genomic medicine in the clinic: The future is here. Genetics in Medicine, 15(4), 258–267.

Mardis, E. R. (2008). Next-generation DNA sequencing methods. Annual Review of Genomics and Human Genetics, 9, 387–402.

Mardis, E. R. (2010). The impact of next-generation sequencing technology on genetics. Trends in Genetics, 26(3), 153–162.

Mardis, E. R. (2012). Genome sequencing and cancer. Current Opinion in Genetics & Development, 22(3), 245–250.

Mardis, E. R. (2013). Next-generation sequencing platforms. Annual Review of Analytical Chemistry, 6, 287–303.

Margulies, M., Egholm, M., Altman, W. E., Attiya, S., Bader, J. S., Bemben, L. A., ... Rothberg, J. M. (2005). Genome sequencing in microfabricated high-density picolitre reactors. Nature, 437(7057), 376–380.

Miller, R. R., Montoya, V., Gardy, J. L., Patrick, D. M., & Tang, P. (2019). Metagenomics for pathogen detection in public health. Annual Review of Public Health, 40, 461–478.

Moran, M. A., Reisch, C. R., Kiene, R. P., & Whitman, W. B. (2013). Genomic insights into bacterial DMSP transformations. Annual Review of Marine Science, 5, 309–333.

Navin, N., Kendall, J., Troge, J., Andrews, P., Rodgers, L., McIndoo, J., ... Hicks, J. (2011). Tumour evolution inferred by single-cell sequencing. Nature, 472(7341), 90–94.

Nazareth T, Li N, Marynchenko M, Zhou Z, Chopra P, Signorovitch J, Wu E, Ahmed S, Marvel J, Sasane R. Burden of illness among patients with fragile X syndrome (FXS): a medicaid perspective. Curr Med Res Opin. (2016);32(3), 405–416

Ng, S. B., Buckingham, K. J., Lee, C., Bigham, A. W., Tabor, H. K., Dent, K. M., ... Bamshad, M. J. (2010). Exome sequencing identifies the cause of a Mendelian disorder. Nature Genetics, 42(1), 30–35.

Patel, A. P., Tirosh, I., Trombetta, J. J., Shalek, A. K., Gillespie, S. M., Wakimoto, H., ... Bernstein, B. E. (2014). Single-cell RNA-seq highlights intratumoral heterogeneity in primary glioblastoma. Science, 344(6190), 1396–1401.

Poretsky, R., Rodriguez-R, L. M., Luo, C., Tsementzi, D., & Konstantinidis, K. T. (2014). Strengths and limitations of 16S rRNA gene amplicon sequencing in revealing temporal microbial community dynamics. PloS One, 9(4), e93827.

Qin, J., Li, R., Raes, J., Arumugam, M., Burgdorf, K. S., Manichanh, C., ... Wang, J. (2010). A human gut microbial gene catalogue established by metagenomic sequencing. Nature, 464(7285), 59–65.

Quick, J., Loman, N. J., Duraffour, S., Simpson, J. T., Severi, E., Cowley, L., ... Carroll, M. W. (2016). Real-time, portable genome sequencing for Ebola surveillance. Nature, 530(7589), 228–232.

Ropers, H. H. (2010). Single gene disorders come into focus again. Nature Reviews Genetics, 11(4), 255–265.

Rothberg, J. M., Hinz, W., Rearick, T. M., Schultz, J., Mileski, W., Davey, M., ... Bustillo, J. (2011). An integrated semiconductor device enabling non-optical genome sequencing. Nature, 475(7356), 348–352.

Schloss, P. D., Gevers, D., & Westcott, S. L. (2011). Reducing the effects of PCR amplification and sequencing artifacts on 16S rRNA-based studies. PloS One, 6(12), e27310.

Shendure, J., Findlay, G. M., & Snyder, M. W. (2017). Genomic medicine–progress, pitfalls, and promise. Cell, 169(1), 11–23.

Sinclair, L., Osman, O. A., Bertilsson, S., & Eiler, A. (2015). Microbial community composition and diversity via 16S rRNA gene amplicons: Evaluating the Illumina platform. PloS One, 10(2), e0116955.

Tang, F., Barbacioru, C., Wang, Y., Nordman, E., Lee, C., Xu, N., ... Surani, M. A. (2009). mRNA-Seq whole-transcriptome analysis of a single cell. Nature Methods, 6(5), 377–382.

The 1000 Genomes Project Consortium. (2015). A global reference for human genetic variation. Nature, 526(7571), 68–74.

Thompson, L. R., et al. (2017). A communal catalogue reveals Earth's multiscale microbial diversity. Nature, 551(7681), 457–463.

Tyson, G. W., Chapman, J., Hugenholtz, P., Allen, E. E., Ram, R. J., Richardson, P. M., ... Banfield, J. F. (2004). Community structure and metabolism through reconstruction of microbial genomes from the environment. Nature, 428(6978), 37–43.

Venter, J. C., Remington, K., Heidelberg, J. F., Halpern, A. L., Rusch, D., Eisen, J. A., ... Smith, H. O. (2004). Environmental genome shotgun sequencing of the Sargasso Sea. Science, 304(5667), 66–74.

Wan, J. C. M., et al. (2017). Liquid biopsies come of age: towards implementation of circulating tumour DNA. Nature Reviews Cancer, 17(4), 223–238.

Weinstein, J. N., et al. (2013). The Cancer Genome Atlas Pan-Cancer analysis project. Nature Genetics, 45(10), 1113–1120.

Wolf, S. M., Lawrenz, F. P., Nelson, C. A., Kahn, J. P., Cho, M. K., Clayton, E. W., ... Wilfond, B. S. (2008). Managing incidental findings in human subjects research: Analysis and recommendations. The Journal of Law, Medicine & Ethics, 36(2), 219–248.

Chapter 6

Nanotechnology-based molecular diagnostics

*Hani Alothaid**

Department of Basic Medical Sciences, Faculty of Medical Sciences, Al-Baha University, Al-Baha, Saudi Arabia

INTRODUCTION

Advancements in nanotechnology-based diagnostics have significantly transformed the domain of molecular diagnostics, offering tools that exhibit superior sensitivity, specificity, and rapidity. These innovations are crucial for the early detection of diseases, the development of personalized medicine, and improvements in point-of-care (POC) testing, ultimately leading to enhanced patient outcomes. This chapter explores recent developments, applications, and prospective trajectories of nanotechnology in diagnostics.

Nanotechnology's impact on molecular diagnostics is profound, primarily due to the unique properties of nanomaterials, such as their large surface area, quantum effects, and ability to be functionalized with various biomolecules. These characteristics enable the development of diagnostic tools that can detect disease markers at extremely low concentrations, which is critical for early disease detection and monitoring (Syedmoradi et al., 2017; Chen et al., 2022).

Nanomaterials, such as quantum dots (QDs), gold nanoparticles, and carbon nanotubes (CNTs), provide exceptional sensitivity and specificity in the identification of biomarkers. For example, QDs possess size-tunable emission characteristics that can be utilized for multiplexed imaging and biosensing, allowing for the concurrent detection of multiple biomarkers (Gao et al., 2019). Gold nanoparticles are widely utilized in colorimetric assays due to their distinctive optical properties, which improve the detection thresholds of various diagnostic tests (Jackson et al., 2017). Additionally, CNTs and graphene, recognized for their remarkable electrical properties, are used in the development of highly sensitive biosensors aimed at detecting proteins, nucleic acids, and other molecules (Bardhan et al., 2020).

The early identification of diseases represents a critical benefit of diagnostics that utilize nanotechnology. Detecting illnesses at an initial stage can greatly enhance treatment efficacy and lower mortality rates (**Figure 6.1**). For instance, assays powered by nanotechnology are capable of identifying cancer biomarkers at minimal concentrations, thereby promoting timely diagnosis and intervention (Tovar-Lopez, 2023). In a similar vein, the incorporation of nanomaterials in POC testing for infectious diseases such as HIV, malaria, and tuberculosis enables swift and precise diagnoses, even in settings with limited resources (Mousavi et al., 2022).

Personalized medicine, which customizes healthcare interventions to align with the unique attributes of individual patients, has seen substantial advancements through the application of nanotechnology-based diagnostic methods. These diagnostic approaches yield intricate molecular data that inform the choice of targeted therapeutic strategies, enhancing treatment effectiveness while reducing the likelihood of adverse reactions. For example, sensors utilizing

* Corresponding author.

DOI: 10.1201/9781003489177-6

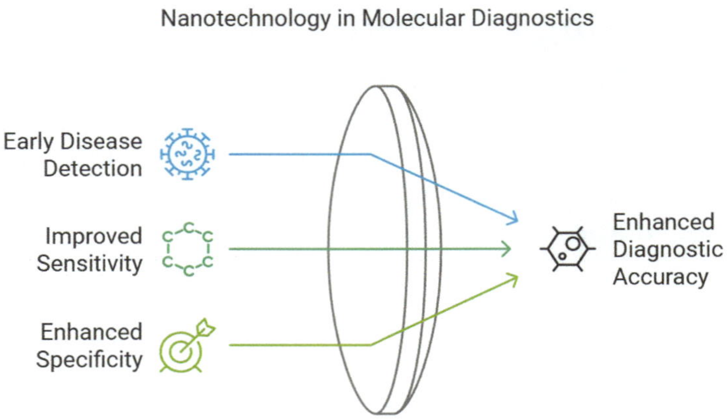

Figure 6.1 Nanotechnology in molecular diagnostics.

nanomaterials can continuously track the concentrations of particular biomarkers, thereby providing critical understanding of a patient's treatment response (Jin & Park, 2019).

The introduction of nanotechnology has significantly transformed POC testing, offering swift, precise, and portable diagnostic solutions that can be utilized at the bedside or in distant settings. These diagnostic tools play a vital role in the management of infectious diseases and chronic illnesses, where prompt diagnosis is critical. Tests that leverage nanotechnology for POC applications are engineered to be accessible and economical, providing results in a matter of minutes and facilitating immediate clinical decision-making (Klostranec et al., 2007; Zhang et al., 2014).

TYPES OF NANOMATERIALS IN DIAGNOSTICS

Quantum dots

QDs are nanoscale semiconductor crystals that possess distinctive optical characteristics, including the ability to emit light at varying wavelengths depending on their size and exceptional resistance to photodegradation (**Table 6.1**). These attributes render QDs highly

Table 6.1 Types of nanomaterials in diagnostics

Nanomaterial	*Unique properties*	*Specific applications*
Quantum Dots (QDs)	Size-tunable light emission, high photostability	Multiplexed imaging, biosensing, cancer biomarker detection
Gold Nanoparticles (AuNPs)	Localized surface plasmon resonance (LSPR), biocompatibility	Colorimetric assays, surface-enhanced Raman scattering (SERS), lateral flow assays
Carbon Nanotubes (CNTs)	Exceptional electrical properties, high surface area	Biosensors for proteins, nucleic acids, small molecules
Graphene	High electrical conductivity, flexibility	Electrochemical sensors, field-effect transistors (GFETs)
Magnetic Nanoparticles (MNPs)	Magnetic properties, ease of functionalization	Magnetic resonance imaging (MRI), magnetoresistive sensors, magnetic relaxation switches

advantageous for applications in multiplexed imaging and biosensing. Recent research has highlighted their efficacy in identifying biomarkers associated with conditions such as cancer and infectious diseases.

QDs exhibit a wide absorption spectrum alongside a narrow and symmetric emission spectrum, which can be precisely adjusted through variations in their size or composition. This ability to tune their properties facilitates the concurrent detection of various targets with a single light source, rendering QDs particularly advantageous for multiplexed assays (Gao et al., 2019). Furthermore, their exceptional photostability enables prolonged imaging without considerable degradation of signal intensity, an essential feature for applications requiring continuous monitoring (Chen et al., 2022).

Recent developments have significantly improved the capabilities of QDs (**Figure 6.2**). Researchers have created conjugated polymer nanoparticles utilizing QDs that exhibit high sensitivity and substantial Stokes shifts for the detection of ferric ions, highlighting their potential utility in biosensing applications (Chen et al., 2022). Furthermore, the integration of QDs with microfluidic systems and advanced signal processing methods has demonstrated considerable promise for high-throughput diagnostics of infectious diseases (Klostranec et al., 2007).

In the field of cancer diagnostics, QDs have been employed for the simultaneous detection of multiple cancer biomarkers. This multiplexing capability is especially advantageous in oncology, as the identification of several biomarkers can provide a more comprehensive understanding of the disease's condition and guide personalized treatment strategies (Mousavi et al., 2022). For instance, QDs have been used in immunoassays to achieve sensitive detection of the hepatitis B surface antigen (HBsAg) and to enable the multiplexed identification of various human viral infections (Zhang et al., 2014). The integration of QDs with other nanomaterials and advanced technologies, such as artificial intelligence and

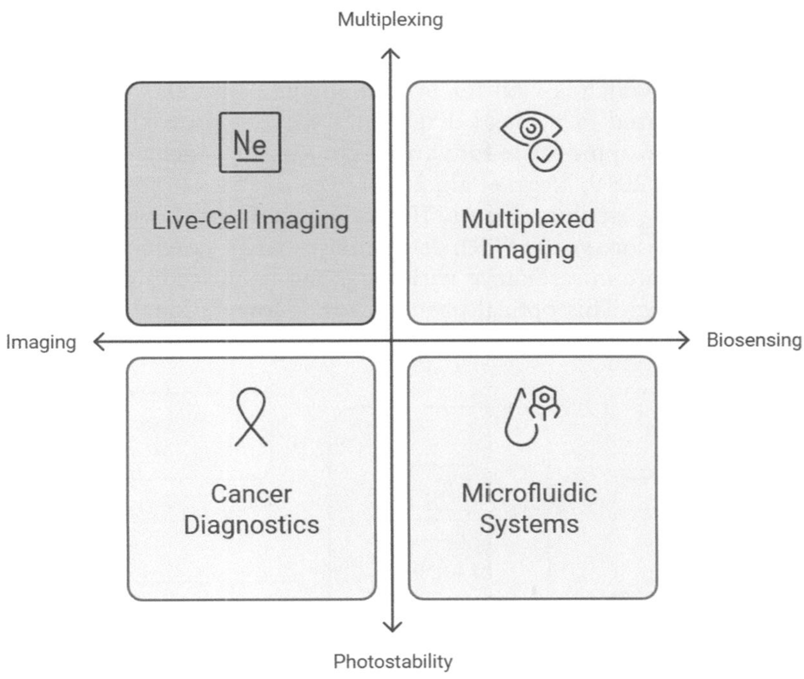

Figure 6.2 Quantum dots in diagnostics and imaging.

machine learning, is expected to further enhance their diagnostic capabilities, paving the way for the creation of more sophisticated and accurate diagnostic tools (Jin & Park, 2019).

QDs are gaining attention for their promising applications in live-cell imaging and in vivo studies. Their capacity to emit bright and stable fluorescence for prolonged durations enables researchers to monitor cellular processes in real-time. This feature is crucial for investigating disease mechanisms and assessing the effectiveness of therapeutic treatments. Furthermore, QDs can be modified with different biological agents, including antibodies, peptides, and nucleic acids, to selectively target specific cells or biomarkers, thereby improving their specificity and minimizing off-target effects (Gao et al., 2019; Mousavi et al., 2022).

Although QDs offer numerous benefits, their application in clinical settings is not without challenges. A significant issue is their possible toxicity, largely attributed to the inclusion of heavy metals like cadmium in some QD formulations. Research is progressing toward the creation of biocompatible and non-toxic alternatives, including QDs derived from carbon or silicon, to address these toxicity concerns (Chen et al., 2022). Furthermore, ensuring the stability and reproducibility of QDs within intricate biological environments continues to be a focus of ongoing research.

Gold nanoparticles

Gold nanoparticles (AuNPs) are extensively employed in diagnostic applications owing to their versatile functionalization, biocompatibility, and distinctive optical properties. The straightforward modification of AuNP surfaces with a variety of biomolecules, including antibodies, peptides, and nucleic acids, enhances their sensitivity and specificity in the detection of target analytes. This ability for functionalization enables targeted interaction with disease-associated biomarkers, thereby significantly elevating the precision and dependability of diagnostic testing (Jin & Park, 2019; Chen et al., 2022). Another significant advantage of gold nanoparticles is their biocompatibility, which promotes their extensive use in clinical diagnostics (**Figure 6.3**). Generally regarded as non-toxic, AuNPs can be safely integrated into biological systems, which is vital for in vivo applications. This characteristic ensures that AuNPs can be utilized in medical diagnostics without provoking adverse biological reactions, rendering them appropriate for various clinical uses, including imaging and POC diagnostics (Jin & Park, 2019; Negri et al., 2020).

Among the remarkable attributes of AuNPs is their optical behavior, particularly localized surface plasmon resonance (LSPR). It manifests when conduction electrons at the surface of AuNPs oscillate in resonance with incoming light, leading to pronounced light absorption and scattering. This optical phenomenon is foundational to several diagnostic

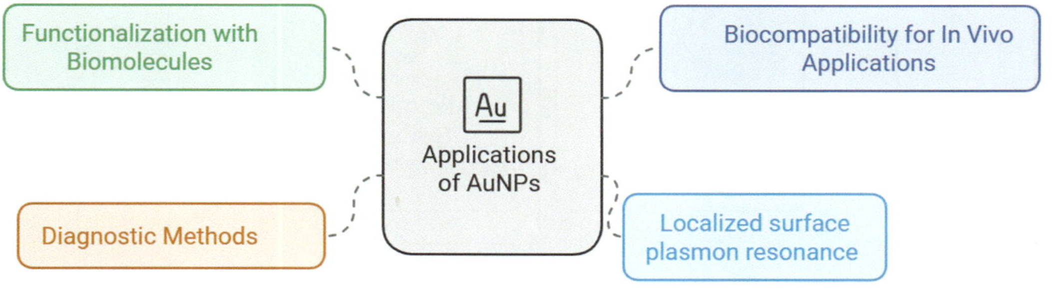

Figure 6.3 Applications of gold nanoparticles (AuNPs).

methodologies, including colorimetric assays, surface-enhanced Raman scattering (SERS) sensors, and lateral flow assays (LFAs). For instance, in colorimetric assays, the binding of AuNPs to a target analyte induces aggregation, shifting the LSPR peak and resulting in a detectable color change, observable either visually or through spectrophotometric methods. This characteristic is harnessed in various diagnostic applications, such as the swift identification of malaria and tuberculosis biomarkers (Mousavi et al., 2022; Jin & Park, 2019).

The remarkable sensitivity and specificity associated with gold nanoparticle (AuNP)-based diagnostic approaches stem from the meticulous engineering and functionalization of these nanoparticles. By precisely manipulating parameters such as size, shape, and surface chemistry, researchers are able to enhance the interactions between AuNPs and their target analytes, thereby improving the efficacy of diagnostic assays. This level of control has resulted in notable enhancements in detection limits and overall accuracy across a range of diagnostic tests, establishing AuNPs as essential elements in contemporary diagnostic methodologies (Jin & Park, 2019; Chen et al., 2022). Looking ahead, advancements in AuNP-based diagnostics are anticipated to prioritize the further enhancement of their sensitivity, specificity, and ability to perform multiplex testing. The integration of AuNPs with cutting-edge technologies, including microfluidics and artificial intelligence, is expected to facilitate the creation of next-generation diagnostic platforms. Such innovations could provide improved precision, reduced turnaround times, and the capability to simultaneously detect multiple biomarkers. It will be essential to overcome the challenges associated with the scalability and reproducibility of AuNP synthesis to enable the widespread implementation of these technologies in clinical practice (Jin & Park, 2019; Chen et al., 2022).

Carbon nanotubes

CNTs and graphene exhibit remarkable electrical and mechanical characteristics, rendering them highly suitable for the development of nanosensors. These materials have found extensive application in biosensors aimed at detecting a diverse array of analytes, such as proteins, nucleic acids, and small molecules. Their incorporation into diagnostic instruments has resulted in notable enhancements in both sensitivity and detection thresholds of these sensors. The outstanding electrical attributes of CNTs and graphene arise from their distinctive structural configurations and substantial surface area. CNTs are cylindrical nanostructures formed from rolled graphene sheets, whereas graphene is characterized as a single layer of carbon atoms organized in a two-dimensional honeycomb lattice. Such structural features contribute to a high surface-to-volume ratio, which promotes the adsorption of biomolecules and improves the interaction between target analytes and the sensor interface (Bardhan et al., 2020; Negri et al., 2020). In the realm of biosensing, CNTs and graphene are frequently modified with specific biomolecules, including antibodies, DNA, or enzymes, to enhance their selectivity toward target analytes. This modification enables the accurate detection of a wide variety of biological molecules. For example, sensors based on CNTs have been engineered to identify proteins linked to cancer and cardiovascular conditions, while graphene-based sensors have been employed to detect nucleic acids from various pathogens (Chen et al., 2022; Mousavi et al., 2022).

The remarkable electrical conductivity exhibited by CNTs and graphene greatly enhances the efficacy of electrochemical sensors. The incorporation of these nanomaterials as transducers in electrochemical biosensors leads to a significant improvement in electron transfer dynamics between the analyte and the electrode. This enhancement translates into heightened sensitivity and reduced detection thresholds, which is particularly advantageous for the identification of low-concentration biomarkers in clinical samples; such capability is essential for the timely diagnosis and surveillance of diseases (Jin & Park, 2019).

Additionally, the exceptional mechanical strength and adaptability of CNTs and graphene facilitate the creation of robust and multifunctional diagnostic tools. These attributes make it possible to design flexible and wearable sensors capable of ongoing, real-time monitoring of physiological metrics. Such innovations are invaluable for the management of chronic ailments and for advancements in personalized medicine (Negri et al., 2020). Moreover, the distinct optical characteristics of graphene further enhance its role in biosensing applications. Graphene-based field-effect transistors (GFETs) have been effectively employed for the detection of various analytes by monitoring alterations in the electrical conductivity of the graphene layer upon the attachment of target molecules. This approach has demonstrated considerable potential in the detection of biomarkers associated with conditions such as diabetes and infectious diseases (Mousavi et al., 2022).

The incorporation of CNTs and graphene into microfluidic devices significantly improves the operational capabilities of diagnostic platforms. Microfluidics facilitates the control of minute fluid volumes, paving the way for the creation of lab-on-a-chip systems capable of executing intricate analyses with limited sample sizes. The synergy of CNTs and graphene within these systems results in enhanced throughput and exceptional sensitivity for diagnostic applications, which are crucial for POC testing and expedited diagnostics (Chen et al., 2022; Bardhan et al., 2020).

MECHANISMS OF NANOTECHNOLOGY-BASED DIAGNOSTICSRC

Optical detection

Optical detection techniques leverage the distinctive optical characteristics of nanomaterials, including fluorescence, plasmon resonance, and chemiluminescence, to identify specific analytes. QDs and gold nanoparticles (AuNPs) are frequently employed due to their exceptional sensitivity and the capacity to be adjusted to particular wavelengths. These techniques are particularly beneficial in imaging applications and for identifying low-abundance biomarkers (Gao et al., 2019; Chen et al., 2022).

QDs are semiconductor nanocrystals characterized by their size-dependent emission properties. Their extensive absorption range and narrow, symmetric emission spectra render them suitable for multiplexed imaging and biosensing applications. QDs can be stimulated by a single light source to emit light across multiple wavelengths, which correspond to various particle sizes, thereby facilitating the concurrent detection of multiple targets (Mousavi et al., 2022). This capability is especially valuable in scenarios requiring the simultaneous identification of multiple biomarkers, such as in cancer diagnostics and the testing of infectious diseases (Klostranec et al., 2007).

Gold nanoparticles (AuNPs) are employed in optical detection techniques primarily due to their localized surface plasmon resonance (LSPR). Upon illumination, the conduction electrons within AuNPs oscillate in unison, leading to pronounced light absorption and scattering at designated wavelengths. This characteristic is particularly useful in colorimetric assays, where alterations in the solution's color signify the presence of a specific analyte (Chang et al., 2019). Additionally, AuNPs are integral to SERS sensors, where the Raman signal of molecules bound to the nanoparticle surface is markedly amplified, facilitating the detection of analytes at extremely low concentrations (Chen et al., 2022).

Another significant detection method is fluorescence-based detection, which frequently utilizes nanomaterials such as QDs and fluorescent dyes (Figure 6.4). These materials emit light when excited, and the intensity of the emitted light is directly proportional to the concentration of the target analyte. This approach is characterized by its high sensitivity and

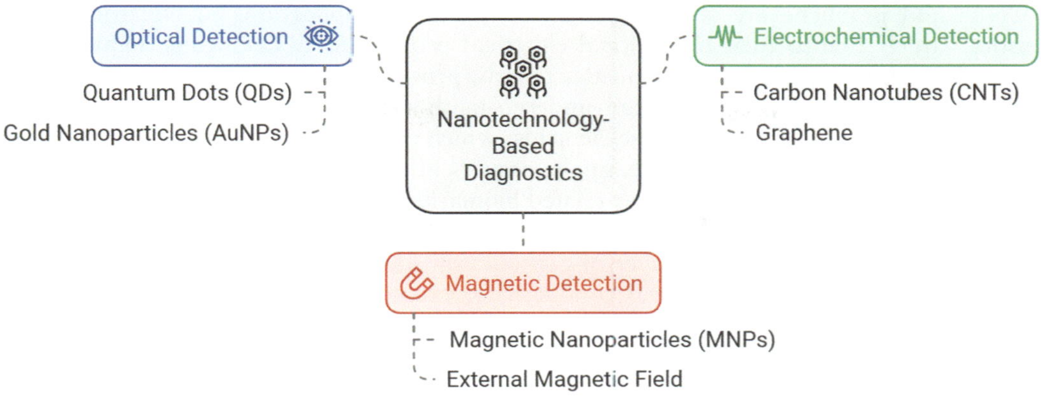

Figure 6.4 Nanotechnology-based diagnostics mechanisms.

is applicable for the real-time detection of various biomarkers, including nucleic acids and proteins (Gao et al., 2019).

Chemiluminescence detection utilizes nanomaterials that produce light through chemical reactions. This technique exhibits high sensitivity, making it suitable for identifying low-abundance biomarkers without relying on external light sources. Assays employing nanoparticles that are functionalized with chemiluminescent probes can effectively detect specific analytes, thereby offering a straightforward and efficient diagnostic approach (Jin & Park, 2019). The combination of optical detection techniques with microfluidic devices has significantly improved the functionality of diagnostics based on nanotechnology. Microfluidics facilitates the handling of minute fluid volumes, leading to the creation of lab-on-a-chip systems capable of executing intricate analyses with limited sample sizes. When integrated with the optical characteristics of nanomaterials, these systems provide high-throughput and exceptionally sensitive diagnostic capabilities, which are crucial for POC testing and expedited diagnostics (Bardhan et al., 2020; Negri et al., 2020).

Electrochemical detection

Electrochemical detection serves as a fundamental element in diagnostics that utilize nanotechnology, capitalizing on the distinctive electrical characteristics of nanomaterials to facilitate the detection of analytes with high sensitivity and specificity. This technique entails the assessment of electrical signals produced by redox reactions or variations in conductivity that occur when target molecules engage with electrodes composed of nanomaterials. Commonly utilized nanomaterials in electrochemical sensors include CNTs, graphene, gold nanoparticles (AuNPs), and metal oxide nanoparticles, all of which are favored for their superior electrical conductivity, substantial surface area, and compatibility with biological systems (Negri et al., 2020; Mousavi et al., 2022). A significant benefit of electrochemical detection lies in its capacity to deliver real-time and quantitative evaluations of a diverse array of analytes, encompassing proteins, nucleic acids, small molecules, and ions. The extensive surface area of nanomaterials amplifies the interaction between the target analytes and the sensor interface, resulting in heightened sensitivity and reduced detection thresholds. This feature is particularly advantageous for identifying low-abundance biomarkers in clinical specimens, which is essential for the early diagnosis and monitoring of diseases (Gao et al., 2019; Chen et al., 2022).

CNTs and graphene are widely recognized for their applications in electrochemical sensors, largely due to their exceptional electrical properties. CNTs, which exhibit high conductivity and impressive mechanical strength, provide a robust platform for developing biosensing technologies. They can undergo modifications with various biomolecules, such as enzymes, antibodies, and nucleic acids, which enhance the selectivity and sensitivity of electrochemical sensors. For example, sensors utilizing CNTs have been designed to detect glucose, cholesterol, and cancer-related biomarkers, highlighting their versatility and effectiveness in clinical diagnostic applications (Bardhan et al., 2020; Negri et al., 2020). Graphene, consisting of a single layer of carbon atoms arranged in a hexagonal lattice, is notable for its exceptional electrical conductivity, extensive surface area, and significant flexibility. These properties make graphene an advantageous material for creating electrochemical sensors. GFETs have been utilized to identify various analytes by tracking changes in the electrical conductivity of the graphene layer upon binding with target molecules. This cutting-edge technology has shown considerable promise in detecting biomarkers related to diseases such as diabetes and infectious illnesses, thus providing rapid and accurate diagnostic results (Chen et al., 2022; Mousavi et al., 2022).

Gold nanoparticles (AuNPs) are widely utilized in the development of electrochemical sensors owing to their distinctive characteristics, such as superior conductivity, ease of surface modification, and compatibility with biological systems. Their presence facilitates improved electron transfer between the target analyte and the sensor electrode, leading to notable enhancements in the sensitivity of electrochemical readings. Applications of AuNPs extend across a range of fields, including the identification of biomolecules like DNA and proteins, as well as small chemical entities. Electrochemical sensors incorporating AuNPs have exhibited remarkable sensitivity and specificity in the detection of biomarkers pertinent to cancer, cardiovascular conditions, and infectious diseases (Jin et al., 2019; Jin & Park, 2019). Similarly, metal oxide nanoparticles, including zinc oxide (ZnO) and titanium dioxide (TiO_2), are employed in electrochemical sensing owing to their high surface area, exceptional catalytic abilities, and biocompatibility. These nanoparticles enhance the capabilities of electrochemical sensors when integrated into their design. For example, ZnO nanorods have facilitated the development of highly sensitive biosensors for glucose monitoring, noted for their rapid response rates. Likewise, TiO_2 nanoparticles have been harnessed in sensors aimed at the detection of hydrogen peroxide and various other analytes, indicating their significant potential in numerous diagnostic applications (Bardhan et al., 2020).

Electrochemical sensors are characterized not only by their exceptional sensitivity and specificity but also by a range of additional benefits. These include affordability, ease of use, and the potential for miniaturization and integration into compact, portable devices. Such attributes render electrochemical detection a compelling choice for POC diagnostics and personalized healthcare. Furthermore, the combination of electrochemical sensors with microfluidic systems and wearable technologies significantly broadens their application scope, facilitating real-time continuous monitoring of physiological parameters (Negri et al., 2020; Chen et al., 2022).

Magnetic detection

Magnetic detection represents a significant advancement in diagnostics rooted in nanotechnology, utilizing the inherent magnetic characteristics of nanoparticles to discern and assess biomolecules with exceptional sensitivity and specificity. This approach employs magnetic nanoparticles (MNPs), which are subject to manipulation and detection through external magnetic fields. Typically constituted of iron oxide (Fe_3O_4 or γ-Fe_2O_3), these nanoparticles can be modified with a variety of biomolecular entities, including antibodies, enzymes, or nucleic acids, to effectively target specific analytes (Gupta & Gupta, 2005; Wahajuddin & Arora, 2012). A key benefit of magnetic detection lies in its capacity to facilitate swift, highly sensitive assays while requiring minimal sample preparation. The external magnetic field allows for the straightforward separation of MNPs from complex biological matrices, thereby enhancing the efficiency of target molecule capture and concentration. This feature is particularly advantageous for applications in POC diagnostics and environmental monitoring, where the demand for prompt and dependable results is critical (Leslie-Pelecky & Rieke, 1996; Pankhurst et al., 2003).

In biosensing domains, MNPs are employed alongside various detection techniques such as magnetic resonance imaging (MRI), magnetoresistive sensors, and magnetic relaxation switches. MRI harnesses the potent magnetic signals produced by MNPs to generate high-resolution images of biological tissues. MNPs that have been functionalized may serve as contrast agents to emphasize particular areas of concern, including tumor locations or inflamed regions, thereby providing critical insights for disease diagnosis and monitoring (Pankhurst et al., 2003; Wahajuddin & Arora, 2012).

Magnetoresistive sensors, such as giant magnetoresistance and tunnel magnetoresistance sensors, detect alterations in electrical resistance induced by a magnetic field. When MNPs attach to a specific analyte, the magnetic field in the vicinity changes, which is subsequently identified by the sensor. This method offers substantial sensitivity and is proficient in identifying low concentrations of biomarkers, which is advantageous for the preliminary diagnosis and surveillance of diseases (Gupta & Gupta, 2005; Gao et al., 2019). Magnetic relaxation switches exploit the variations in the relaxation time of MNPs that occur upon the binding of target molecules. When MNPs are functionalized with distinct ligands, they engage with target analytes, resulting in modifications to their magnetic relaxation characteristics. These alterations can be assessed through nuclear magnetic resonance or other magnetic detection methods, yielding a quantitative evaluation of the analyte's concentration (Leslie-Pelecky & Rieke, 1996; Pankhurst et al., 2003). The amalgamation of magnetic detection techniques with microfluidic devices has further elevated the potential of diagnostics rooted in nanotechnology. Microfluidic platforms facilitate meticulous management and manipulation of minuscule volumes of fluids, enabling the creation of lab-on-a-chip systems that integrate sample processing, magnetic separation, and detection within a singular apparatus. These systems provide elevated throughput and heightened sensitivity in diagnostic applications, which are critical for advanced clinical assessments.

Applications in disease detection

Advancements in diagnostics utilizing nanotechnology represent a transformative approach in disease detection, allowing for the highly sensitive and precise identification of ailments in their nascent stages. These innovative diagnostic methods employ nanomaterials, including QDs, silica nanospheres, and MNPs, to augment biomolecule detection, refine imaging methodologies, and enable POC assessments (see **Table 6.2**). For example, QDs are employed due to their exceptional optical characteristics in numerous diagnostic applications, while MNPs play a crucial role in isolating and identifying rare cells and pathogens within various samples.

Cancer diagnostics

Nanotechnology-driven diagnostic methods have made notable advancements in the early detection and management of cancer. Innovations such as nanoparticle-enhanced imaging and assays utilizing nanomaterials have been established for the identification of cancer

Table 6.2 Applications of nanotechnology-based diagnostics in disease detection

Disease	*Nanomaterial used*	*Diagnostic method*	*Benefits*
Cancer	Quantum Dots, Gold Nanoparticles, Carbon Nanotubes	Multiplexed imaging, SERS, colorimetric assays	Early detection, high sensitivity, comprehensive biomarker profiling
Infectious diseases	Gold Nanoparticles, Magnetic Nanoparticles, Quantum Dots	Lateral flow assays, magnetic separation, fluorescence-based assays	Rapid diagnosis, point-of-care testing, high specificity
Neurological disorders	Gold Nanoparticles, Quantum Dots, Magnetic Nanoparticles	Colorimetric assays, fluorescence-based detection, MRI	Early diagnosis, real-time monitoring, non-invasive methods

biomarkers present in minuscule quantities. These technological improvements facilitate the detection of cancer at a more initial phase, which is vital for favorable treatment results (Chen et al., 2022; Gao et al., 2019). The imaging techniques enhanced by nanoparticles exploit the distinctive optical and magnetic traits of these particles to augment the contrast and resolution of imaging techniques such as MRI, computed tomography (CT), and positron emission tomography. For example, gold nanoparticles (AuNPs) and iron oxide nanoparticles are frequently employed as contrast agents in these imaging methodologies. Their extensive surface area and ease of modification allow for the targeted identification of cancerous cells, thereby improving the visualization of tumors and metastatic sites (Jin & Park, 2019; Wahajuddin & Arora, 2012).

Gold nanoparticles that are conjugated with antibodies specific to the HER2 receptor have been employed to enhance the imaging of breast cancer lesions. The HER2 receptor is frequently overexpressed in certain forms of breast cancer, which makes it a suitable target for diagnostic applications. Research indicated that these gold nanoparticles (AuNPs) markedly enhanced the contrast in computed tomography (CT) scans, facilitating the clear identification of HER2-positive tumors at early developmental stages (Jin & Park, 2019). This method not only aided in early tumor detection but also enabled the monitoring of the tumor's response to therapeutic interventions, thus serving as a comprehensive instrument for cancer management. Nanomaterial-based assays harness the remarkable sensitivity and specificity of various nanomaterials for the detection of cancer biomarkers within biological specimens, including blood, urine, and tissue samples. QDs, AuNPs, and CNTs are commonly utilized in these assays due to their outstanding optical and electrical properties. For instance, immunoassays that incorporate QDs can simultaneously identify multiple cancer biomarkers, thereby offering a thorough characterization of the disease and assisting in tailored treatment strategies (Pankhurst et al., 2003; Mousavi et al., 2022). A study focusing on QDs illustrated their capability for the concurrent detection of various cancer biomarkers. Researchers established a multiplexed assay utilizing QDs paired with distinct antibodies, each aimed at specific biomarkers such as prostate-specific antigen, carcinoembryonic antigen, and alpha-fetoprotein. This assay enabled the identification of multiple biomarkers in a single test with high sensitivity and specificity, thus significantly enhancing the diagnostic process for different cancers (Mousavi et al., 2022). The multiplexed strategy not only minimizes the time and expense associated with diagnostics but also enriches the understanding of the patient's cancer profile.

SERS represents a significant advancement in cancer diagnostics, facilitated by nanotechnology. This technique employs metallic nanoparticles, predominantly gold or silver, to enhance the Raman signals of specific target molecules, thereby enabling the identification of cancer biomarkers even at minimal concentrations. SERS has demonstrated considerable efficacy in identifying biomarkers associated with various cancers, such as breast, lung, and prostate cancer (Leslie-Pelecky & Rieke, 1996; Chen et al., 2022).

In a notable application, researchers utilized gold nanoparticles to create an SERS-based assay aimed at detecting mutations in the epidermal growth factor receptor (EGFR), which are frequently observed in lung cancer patients. This SERS assay exhibited remarkable sensitivity and specificity, successfully identifying EGFR mutations at very low concentrations in blood samples from patients. This non-invasive diagnostic approach facilitates early detection and ongoing monitoring of lung cancer, offering a substantial advantage over conventional biopsy techniques (Leslie-Pelecky & Rieke, 1996). Additionally, circulating tumor cells (CTCs) and cell-free nucleic acids (cfNAs) found in blood are gaining recognition as important biomarkers for the early detection and monitoring of cancer. The integration of nanotechnology has led to the creation of highly sensitive and specific methodologies for the isolation and analysis of these biomarkers. For example, microfluidic devices that incorporate nanomaterials can effectively capture CTCs from blood samples, allowing for subsequent analysis of genetic mutations and other cancer-specific indicators. Likewise, nanoparticles that are functionalized with DNA probes can be employed to detect cfNAs, offering valuable insights into tumor genetics and supporting the development of targeted therapeutic strategies (Gupta & Gupta, 2005; Gao et al., 2019).

A recent investigation employed a microfluidic device that incorporates graphene oxide nanosheets to effectively capture CTCs from the blood of breast cancer patients. This device successfully isolated CTCs by leveraging their distinct physical and biological characteristics, facilitating subsequent genetic analyses. This methodology allowed for the identification of CTCs at extremely low concentrations, thereby serving as a potent instrument for the early detection of cancer and the assessment of treatment effectiveness (Gao et al., 2019). Beyond its diagnostic capabilities, nanotechnology is pivotal in evaluating treatment responses and identifying cancer recurrence. Nanoparticles can be engineered to deliver therapeutic agents directly to tumor cells while simultaneously offering real-time insights into treatment effectiveness through imaging or biomarker detection. This comprehensive strategy, referred to as theranostics, enables ongoing assessment of treatment progress and the prompt identification of potential recurrences, ultimately enhancing patient outcomes (Pankhurst et al., 2003; Wahajuddin & Arora, 2012). In a notable study, researchers created theranostic nanoparticles featuring iron oxide cores enveloped in a gold layer. These nanoparticles were functionalized with ligands specific to prostate-specific membrane antigen, facilitating targeted delivery to prostate cancer cells. They provided high-contrast MRI imaging for diagnostic purposes while also delivering therapeutic agents directly to the tumor site. This dual functionality allowed for real-time monitoring of treatment effectiveness and the early detection of tumor recurrence, significantly benefiting patient outcomes (Pankhurst et al., 2003).

Infectious diseases

Nanotechnology has significantly contributed to the fight against infectious diseases by offering rapid and precise diagnostic solutions. The integration of nanomaterials has led to the creation of POC testing systems capable of swiftly identifying infections like HIV, tuberculosis (TB), and malaria, even in environments with limited resources. These diagnostic tools are designed to be accessible, economically viable, and proficient in providing results within a matter of minutes (Wahajuddin & Arora, 2012; Mousavi et al., 2022). Nanomaterials such as gold nanoparticles, QDs, and MNPs are integral to these diagnostic technologies. Their unique properties, such as high surface area, ease of functionalization, and exceptional optical and magnetic characteristics, enable the development of highly sensitive and specific diagnostic assays.

Gold nanoparticles (AuNPs) have been effectively employed in LFAs for the swift identification of malaria. These assays utilize AuNPs that are conjugated with antibodies specifically designed to bind to malaria antigens found in blood samples. Upon the application of a blood sample to the test strip, the binding of antigens to antibodies results in the formation of a visible line, signifying a positive outcome. This diagnostic technique offers a rapid and dependable means of diagnosis within minutes, which is particularly vital in remote and resource-constrained environments where laboratory facilities may not be accessible (Jin & Park, 2019; Chen et al., 2022). QDs have been utilized in the creation of POC diagnostic tests for HIV. In a particular study, QDs conjugated with antibodies targeting HIV antigens were employed in a fluorescence-based assay. The detection of HIV antigens in the sample triggered the QDs to emit a fluorescent signal, which could be identified using a portable fluorescence reader. This methodology facilitated the highly sensitive identification of HIV at very low concentrations, serving as a significant tool for the early diagnosis and monitoring of HIV infection (Gao et al., 2019; Mousavi et al., 2022;). MNPs have been incorporated into diagnostic assays for the identification of TB. These nanoparticles can be modified with ligands that specifically attach to TB biomarkers. When a sample containing TB bacteria is introduced, the MNPs capture the bacteria, allowing for the complex to be isolated through the application of an external magnetic field. This technique not only concentrates the target bacteria but also enables rapid detection using magnetic sensors. Consequently, this method considerably decreases the time needed for TB diagnosis in comparison to traditional culture techniques, which may require several weeks (Pankhurst et al., 2003; Wahajuddin & Arora, 2012).

The COVID-19 pandemic underscored the urgency for swift and precise diagnostic methods. Nanotechnology emerged as a crucial element in the creation of POC tests aimed at identifying SARS-CoV-2, the virus that causes COVID-19. A notable illustration of this is the employment of gold nanoparticles in lateral flow immunoassays designed to detect viral antigens in nasopharyngeal swabs. These diagnostic tests delivered results in as little as 15 minutes, facilitating prompt screening and the isolation of infected individuals, which was instrumental in mitigating the transmission of the virus (Chen et al., 2022; Mousavi et al., 2022). Moreover, diagnostic tools based on nanotechnology significantly enhance the sensitivity and specificity of testing methodologies while also allowing for multiplexing, enabling the simultaneous detection of multiple pathogens in a single assay. This feature holds particular importance in resource-constrained environments where the availability of extensive laboratory capabilities is frequently limited. Beyond diagnostic uses, nanotechnology contributes to the fabrication of portable and user-friendly devices suitable for field deployment. These devices often leverage microfluidic technologies, which facilitate the accurate manipulation of minimal sample and reagent volumes, thereby improving the overall efficiency and precision of diagnostic assays (Leslie-Pelecky & Rieke, 1996; Gao et al., 2019).

NEUROLOGICAL DISORDERS

The utilization of nanotechnology in the diagnosis of neurological disorders, including Alzheimer's disease and Parkinson's disease, has yielded encouraging outcomes. Nanoparticles have been employed to identify specific biomarkers within biological fluids, thus enabling early detection and ongoing assessment of these diseases. Recent advancements have emphasized the creation of nanomaterial-based sensors capable of penetrating the blood-brain barrier (BBB), offering the potential for real-time observation of disease progression (Hwang et al., 2018; Xie et al., 2020). A primary obstacle in the diagnosis of neurological disorders is the identification of biomarkers that are frequently found in extremely low concentrations within cerebrospinal fluid or blood. Various nanoparticles, including gold nanoparticles (AuNPs), QDs, and MNPs, have been utilized to improve the sensitivity and specificity of biomarker detection. These nanomaterials can be modified with antibodies, peptides, or nucleic acids to selectively target biomarkers related to neurodegenerative conditions (Huang et al., 2019; Zhao et al., 2020). Gold nanoparticles have been instrumental in creating ultra-sensitive assays for the identification of amyloid-beta (Aβ) peptides, which serve as critical biomarkers for Alzheimer's disease. In one investigation, a colorimetric assay was developed using AuNPs modified with anti-Aβ antibodies. The interaction of Aβ peptides in the sample prompted the aggregation of AuNPs, resulting in a noticeable color change that could be quantified with ease. This assay enabled the detection of Aβ at picomolar levels, greatly enhancing the prospects for early diagnosis of Alzheimer's disease (Huang et al., 2019; Zhao et al., 2020).

QDs have been employed to monitor the progression of Parkinson's disease by the identification of alpha-synuclein aggregates within the brain. A significant study demonstrated the conjugation of QDs with specific ligands that selectively bind to alpha-synuclein. The fluorescence emitted by these QDs upon interaction with the aggregates allowed for the visualization and quantification of these biomarkers in brain tissues. This innovative approach enabled real-time assessment of disease progression and the evaluation of therapeutic efficacy (Xie et al., 2020). MNPs have emerged as promising agents for traversing the BBB to deliver diagnostic substances directly to cerebral tissues. In a pioneering strategy, MNPs were modified with transferrin, a protein capable of crossing the BBB through receptor-mediated transcytosis. This modification facilitated the delivery of imaging agents to the brain, significantly improving MRI detection of early-stage Alzheimer's disease. The capacity to penetrate the BBB and target specific regions of the brain presents new opportunities for non-invasive diagnosis and monitoring of neurological conditions (Pankhurst et al., 2003; Hwang et al., 2018).

Additionally, sensors based on nanotechnology are being developed to enable continuous, real-time monitoring of neurological disorders. For instance, electrochemical sensors incorporating graphene and CNTs have been engineered to detect neurotransmitters and other neurochemicals within the brain. These sensors can be implanted to facilitate ongoing observation of chemical fluctuations associated with neurodegenerative diseases, thereby providing critical insights into the underlying mechanisms and progression of these conditions (Huang et al., 2019; Zhao et al., 2020). Nanotechnology not only enhances diagnostic and monitoring capabilities but also offers significant potential for the treatment of neurological disorders. By engineering nanoparticles to transport therapeutic agents directly to targeted areas of the brain, this method may improve treatment effectiveness while minimizing adverse effects. This innovative strategy, referred to as theranostics, integrates both diagnostic and therapeutic functions within a single nanomaterial framework, thereby presenting a holistic approach to the management of neurological conditions (Pankhurst et al., 2003; Xie et al., 2020).

Advances in point-of-care testing

POC testing has experienced significant progress through the incorporation of nanotechnology, which has improved the speed and precision of diagnostics conducted at the patient's location or in remote settings. These tests are engineered to be portable, easy to use, and capable of delivering immediate results, which is essential for the effective management of infectious diseases and chronic illnesses (Wang et al., 2019; Zhou et al., 2021). Nanomaterials, including gold nanoparticles (AuNPs), QDs, and MNPs, play a pivotal role in the creation of sophisticated POC diagnostic instruments. These materials exhibit distinctive characteristics, such as a high surface area, straightforward functionalization, and remarkable optical and magnetic properties, which significantly enhance the sensitivity and specificity of diagnostic tests.

Gold nanoparticles (AuNPs) are extensively employed in LFAs, which represent a prevalent category of POC testing. In these assays, AuNPs are conjugated with antibodies that specifically bind to target antigens present in a sample. The interaction between the antibody and antigen results in a discernible color change, allowing for straightforward interpretation of results without the need for specialized laboratory equipment. This characteristic of simplicity and dependability is particularly crucial in settings with limited resources, where access to laboratory facilities is often restricted (Jin & Park, 2019; Wang et al., 2019). A recent review highlights that AuNPs also provide a range of signal transduction mechanisms, such as plasmonic and photothermal effects, which further enhance their applicability in POC testing (Luciano et al, 2022). QDs significantly improve fluorescence-based POC diagnostics. These semiconductor nanocrystals are characterized by their size-dependent light emission and exceptional photostability, rendering them suitable for multiplexed assays. QDs can be conjugated with antibodies or other ligands, enabling the simultaneous binding of multiple targets and facilitating a rapid and comprehensive diagnostic assessment. This ability to multiplex is especially advantageous for the detection of various biomarkers within a single test (Zhou et al., 2021; Mousavi et al., 2022).

MNPs present novel strategies for POC testing. These nanoparticles are modified with ligands that specifically capture targeted biomarkers. After binding to their respective targets, MNPs can be isolated from the sample using an external magnetic field, which concentrates the analyte for enhanced detection sensitivity. This approach is particularly effective for isolating and identifying pathogens from complex biological matrices such as blood or saliva (Pankhurst et al., 2003; Wahajuddin & Arora, 2012). MNPs enable the efficient detection of bacteria, cells, biomolecules, and proteins due to their superparamagnetic properties and high surface area-to-volume ratio, making them highly versatile for biosensor applications (Wang et al., 2024).

The combination of nanotechnology and microfluidics has significantly transformed POC testing. Microfluidic devices are adept at managing small volumes of fluids, facilitating the creation of lab-on-a-chip systems capable of executing intricate analyses with minimal sample inputs. By integrating the high sensitivity offered by nanomaterials with the precise fluid management characteristic of microfluidics, these systems produce portable, high-throughput diagnostic tools. They are designed to swiftly assess samples, furnishing immediate results that are crucial for timely medical responses (Wang et al., 2019; Zhou et al., 2021). POC tests utilizing nanotechnology present numerous benefits compared to traditional laboratory diagnostics. They are generally more rapid, portable, and require less specialized knowledge for operation. These factors render them particularly advantageous in resource-limited environments and during critical scenarios that demand quick decision-making, such as infectious disease outbreaks. Furthermore, these technological advancements have notable social implications. For example, the application of noble metal nanoparticles in POC testing has enhanced living standards in resource-constrained regions by facilitating swift disease diagnosis and monitoring. These tests harness diverse signal transduction mechanisms, including plasmonics, catalysis, and photothermal effects, which can be finely adjusted and optimized to improve their operational efficiency (Luciano et al, 2022).

CONCLUSION

In summary, nanotechnology has fundamentally transformed molecular diagnostics by facilitating the detection of disease biomarkers with unprecedented sensitivity, specificity, and speed, which are essential for early diagnosis and the advancement of precision medicine. This chapter has explored the distinct characteristics of various nanomaterials, such as QDs, gold nanoparticles, and CNTs, and their contributions to improving diagnostic techniques. Through the application of optical, electrochemical, and magnetic detection methods, diagnostics based on nanotechnology are poised for extensive use in many areas, including oncology, infectious diseases, and POC applications. Nonetheless, despite significant progress, there are still challenges to be addressed, including the need for biocompatibility, safety, and consistency of nanomaterials within clinical environments. Future research should focus on creating environmentally friendly, biocompatible materials and incorporating innovative technologies, such as artificial intelligence, to enhance diagnostic accuracy and scalability. With ongoing advancements, nanotechnology-driven molecular diagnostics has the potential to address critical shortcomings in global healthcare, particularly in resource-limited environments, thus working toward the goal of making diagnostic solutions more accessible, precise, and personalized. The continuous development of this discipline is expected to have a profound impact on the future landscape of medical diagnostics and public health.

REFERENCES

Bardhan, N. M., Jansen, P., & Belcher, A. M. (2020). Graphene, carbon nanotube and plasmonic nanosensors for detection of viral pathogens: Opportunities for rapid testing. Advanced Materials, 32(45), 2002193.

Chang, C.-C., Chen, C.-P., Wu, T.-H., Yang, C.-H., Lin, C.-W., & Chen, C.-Y. (2019). Gold nanoparticle-based colorimetric strategies for chemical and biological sensing applications. Nanomaterials, 9, 861. https://doi.org/10.3390/nano9060861

Chen, X., Hussain, S., Abbas, A., Hao, Y., Malik, A. H., Tian, X., ... Gao, R. (2022). Conjugated polymer nanoparticles and their nanohybrids as smart photoluminescent and photoresponsive material for biosensing, imaging, and theranostics. Microchimica Acta, 189, 83.

Gao, Z. Y., Zhang, X., Xing, S., Lu, Q., Yao, J. S., Liu, Q. Z., ... Ding, B. (2019). conjugated polymer nanoparticles based on carbazole for detecting ferric ion (III) with a large stokes shift and high sensitivity and the application in cell. Journal of Photochemistry and Photobiology A: Chemistry, 386, 112081.

Gupta, A. K., & Gupta, M. (2005). Synthesis and surface engineering of iron oxide nanoparticles for biomedical applications. Biomaterials, 26(18), 3995–4021.

Huang, J., Xie, W., Sun, X., Zhang, Y., & Zhang, X. (2019). Gold nanoparticle-based colorimetric assay for the detection of amyloid-beta peptides. Biosensors and Bioelectronics, 135, 123–130.

Hwang, S., Lee, S., Jun, B. H., & Kim, Y. S. (2018). Multifunctional magnetic nanoparticles for imaging and therapy of Alzheimer's disease. Journal of Nanobiotechnology, 16(1), 35.

Jackson, T. C., Patani, B. O., & Ekpa, D. E. (2017). Nanotechnology in diagnosis: A review. Advanced Nanoparticles, 6(2), 93–102.

Jin, W., & Park, D. H. (2019). functional layered double hydroxide nanohybrids for biomedical imaging. Nanomaterials, 9(10), 1404.

Klostranec, J. M., Xiang, Q., Farcas, G. A., Lee, J. A., Rhee, A., Lafferty, E. I., ... Chan, W. C. W. (2007). Convergence of quantum dot barcodes with microfluidics and signal processing for multiplexed high-throughput infectious disease diagnostics. Nano Letters, 7(9), 2812–2818.

Leslie-Pelecky, D. L., & Rieke, R. D. (1996). Magnetic properties of nanostructured materials. Chemistry of Materials, 8(8), 1770–1783.

Luciano K, Wang X, Liu Y, Eyler G, Qin Z, Xia X. (2022). Noble metal nanoparticles for point-of-care testing: Recent advancements and social impacts. Bioengineering, 9(11), 666.

Mousavi, S. M., Hashemi, S. A., Kalashgrani, M. Y., Omidifar, N., Lai, C. W., Rao, N. V., ... Gholami, A. (2022). The pivotal role of quantum dots-based biomarkers integrated with ultrasensitive probes for multiplex detection of human viral infections. Pharmaceuticals, 15(8), 880.

Negri, V., Pacheco-Torres, J., Calle, D., & López-Larrubia, P. (2020). Carbon nanotubes in biomedicine. Topics in Current Chemistry, 378, 15.

Pankhurst, Q. A., Connolly, J., Jones, S. K., & Dobson, J. (2003). Applications of magnetic nanoparticles in biomedicine. Journal of Physics D: Applied Physics, 36(13), R167–R181.

Syedmoradi, L., Daneshpour, M., Alvandipour, M., Gomez, F. A., Hajghassem, H., & Omidfar, K. (2017). Point of care testing: The impact of nanotechnology. Biosensors and Bioelectronics, 87, 373–387.

Tovar-Lopez, F. J. (2023). Recent progress in micro- and nanotechnology-enabled sensors for biomedical and environmental challenges. Sensors, 23(12), 5406.

Wahajuddin, & Arora, S. (2012). Superparamagnetic iron oxide nanoparticles: magnetic nanoplatforms as drug carriers. International Journal of Nanomedicine, 7, 3445–3471.

Wang M, Jin L, Hang-Mei Leung P, Wang-Ngai Chow F, Zhao X, Chen H, Pan W, Liu H & Li S. (2024) Advancements in magnetic nanoparticle-based biosensors for point-of-care testing. Frontiers in Bioengineering and Biotechnology, 12, 1393789.

Wang, S., Inci, F., De Libero, G., Singhal, A., & Demirci, U. (2019). Point-of-care assays for tuberculosis: Role of nanotechnology/microfluidics. Biotechnology Advances, 37(7), 107–129.

Xie, J., Yang, T., & Ho, D. (2020). Advances in nanoparticles for the diagnosis and treatment of Alzheimer's disease. Advanced Materials, 32(15), 2002066.

Zhang, P., Lu, H., Chen, J., Han, H., & Ma, W. (2014). Simple and sensitive detection of HBsAg by using a quantum dots nanobeads based dot-blot immunoassay. Theranostics, 4(4), 307–315.

Zhao, Y., Gao, S., Zhang, Z., Lin, Y., Hu, W., & Wen, Z. (2020). Quantum dots-based immunoassay for the detection of alpha-synuclein aggregates in Parkinson's disease. Biosensors and Bioelectronics, 147, 111781.

Zhou, L., Wang, L., Xu, Z., & Wei, X. (2021). Recent developments in nanoparticle-based point-of-care diagnostic systems for infectious diseases. Journal of Nanobiotechnology, 19(1), 1–23.

Chapter 7

Role of artificial intelligence in molecular diagnostics

Abdullah Hamadi[*]

Department of Medical Laboratory Technology, Faculty of Applied Medical Sciences, University of Tabuk, Tabuk, Saudi Arabia

Atif Abdulwahab A. Oyouni

Department of Biology, Faculty of Science, University of Tabuk, Tabuk, Saudi Arabia

Osama M. Al-Amer

Department of Medical Laboratory Technology, Faculty of Applied Medical Sciences, University of Tabuk, Tabuk, Saudi Arabia

INTRODUCTION

Artificial intelligence (AI) has transformed the field of molecular diagnostics by markedly improving the accuracy, efficiency, and speed of diagnostic procedures. AI algorithms, especially those based on machine learning (ML) and deep learning techniques, possess the capability to analyze extensive datasets derived from genetic sequences, medical imaging, and electronic health records (EHRs). These algorithms can detect patterns that may be overlooked by human clinicians. The capacity to process and interpret large volumes of data with exceptional precision is essential for the progression of molecular diagnostics (**Table 7.1**).

In the realm of digital pathology, AI platforms such as QuPath and Tissue Phenomics have played a crucial role in the automation of cancerous tissue detection and classification. These technologies utilize both supervised and unsupervised learning methodologies to enhance the precision of pathological evaluations, thus supporting pathologists in delivering more accurate diagnoses. QuPath, which is an open-source application, offers features for the analysis of whole-slide images, tumor detection, and the quantitative assessment of biomarkers through supervised classification techniques (Ponomarev et al., 2017; Jones et al., 2021). Meanwhile, Tissue Phenomics, created by AstraZeneca, has been utilized in clinical initiatives focused on immune oncology, facilitating the identification of essential biomarkers for cancer diagnosis (Smith et al., 2014).

The integration of AI in molecular diagnostics extends beyond digital pathology. In genomics, AI algorithms are essential for analyzing next-generation sequencing (NGS) data, facilitating the detection of genetic mutations and variants associated with different diseases. These algorithms can swiftly process large datasets to pinpoint clinically relevant genetic alterations, which aids in early disease detection and the formulation of personalized treatment approaches (Liu et al., 2019; Wang et al., 2020).

Additionally, AI has been utilized in the examination of EHRs to identify relationships between genetic information and clinical results. This integration facilitates a deeper

[*] Corresponding author.

DOI: 10.1201/9781003489177-7

Table 7.1 Applications of AI in molecular diagnostics

Application area	*Role of AI*	*Benefits*	*Examples*
Digital pathology	Automated detection and classification of tissues	Increased accuracy, faster diagnosis, aid to pathologists	QuPath, Tissue Phenomics
Genomic analysis	Analysis of NGS data, identification of genetic mutations	Early disease detection, personalized treatment plans	AI algorithms for NGS data analysis
Electronic health records (EHRs)	Correlation of genetic data with clinical outcomes	Comprehensive understanding of patient health, informed decision-making	Analysis of EHRs to uncover correlations
Medical imaging	Detection and classification of anomalies in imaging data	Improved accuracy, reduced workload for radiologists	CNNs for lung nodule detection, breast cancer classification

comprehension of patient health, thereby supporting more informed decision-making and enhancing the quality of patient care (Rajkomar et al., 2018).

ML includes a range of algorithms that serve various purposes in diagnostics. These algorithms can be generally divided into two categories: supervised and unsupervised learning methods, each providing distinct benefits in the field of medical diagnostics (**Figure 7.1**).

Supervised learning algorithms, including support vector machines (SVMs) and random forests, are extensively employed for the classification of diseases utilizing labeled datasets. These algorithms acquire knowledge from a training dataset that comprises input-output pairs, enabling them to predict or classify new, previously unseen data. For example, SVMs demonstrate notable efficacy in differentiating between benign and malignant tumors through the analysis of histopathological images (Liu et al., 2019). Additionally, random forests, which consist of multiple decision trees, are applied in various diagnostic applications, such as forecasting patient outcomes and identifying disease biomarkers (Breiman, 2001).

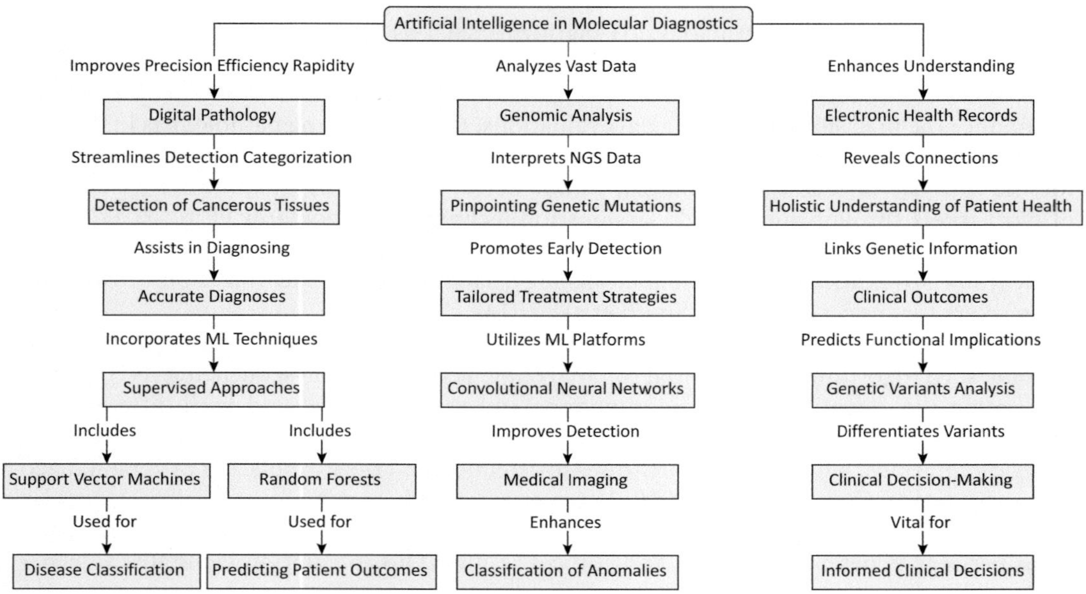

Figure 7.1 Flowchart illustrating the diverse applications of artificial intelligence (AI) in molecular diagnostics.

Unsupervised learning plays a vital role in uncovering concealed patterns within data without the necessity of prior labeling. This methodology is especially beneficial for the detection of emerging diseases and for gaining insights into intricate biological processes. Clustering techniques, such as k-means and hierarchical clustering, are employed to categorize similar patient profiles or genetic sequences, potentially resulting in the identification of novel disease subtypes or variations in treatment responses (Xu & Tian, 2015).

Certain ML architectures, especially convolutional neural networks (CNNs), have revolutionized the field of medical imaging by enhancing the detection and classification of abnormalities. CNNs are particularly adept at recognizing patterns in imaging data due to their ability to comprehend spatial hierarchies. They have been extensively applied in identifying lung nodules in CT scans and classifying breast cancer lesions in mammograms (Litjens et al., 2017). These models are trained on large datasets of labeled images, enabling them to detect subtle differences that may be missed by human radiologists (Esteva et al., 2017).

Recurrent neural networks (RNNs) and long short-term memory (LSTM) networks represent additional ML architectures utilized in diagnostic applications, especially for the analysis of time-series data, including electrocardiograms (ECGs) and EHRs. These models are specifically engineered to identify temporal dependencies within the data, rendering them effective for forecasting disease progression and patient outcomes over time (Shickel et al., 2018).

Additionally, ML algorithms have been employed in the field of genomics to analyze data from NGS. These algorithms play a crucial role in identifying genetic mutations and variants linked to various diseases, thereby enhancing early detection and the development of personalized treatment strategies (Liu et al., 2019). For instance, deep learning models are capable of predicting the functional consequences of genetic variants, which assists in the diagnosis of genetic disorders (Zhou & Troyanskaya, 2015).

ML algorithms improve diagnostic precision and increase efficiency by automating repetitive tasks. For example, natural language processing methods are employed to extract pertinent information from unstructured clinical notes, facilitating faster and more accurate data analysis (Johnson et al., 2017).

The incorporation of ML algorithms in diagnostic processes enhances precision while simultaneously decreasing the time needed for diagnosis, facilitating more rapid medical interventions. As these technologies advance, their utilization in diagnostics is anticipated to broaden, further revolutionizing the healthcare sector.

SPECIFIC ML ALGORITHM IN MEDICAL IMAGING

Certain ML platforms, especially CNNs, have transformed the field of medical imaging by greatly improving the identification and categorization of abnormalities. CNNs excel in pattern recognition within imaging data, owing to their capacity to understand spatial hierarchies, which renders them particularly effective for applications in medical diagnostics (**Table 7.2**).

CNNs are structured with several layers that systematically process input images using convolutional filters, pooling techniques, and fully connected layers. This layered architecture empowers CNNs to learn intricate features and representations. As a result, CNNs are capable of autonomously identifying and classifying irregularities in medical images with remarkable accuracy. For example, these networks have been widely utilized in the detection of lung nodules in CT scans, playing a crucial role in the early identification of lung cancer. The models scrutinize slices of CT scans to pinpoint nodules, which may serve as potential indicators of malignancy, achieving a level of precision that is comparable to or even exceeds that of human radiologists (Litjens et al., 2017; Ardila et al., 2019).

Table 7.2 Machine learning algorithms used in molecular diagnostics

Algorithm	*Type*	*Specific applications*	*Benefits*
Support vector machines (SVMs)	Supervised	Disease classification, distinguishing between benign and malignant tumors	High accuracy in classification tasks
Random forests	Supervised	Predicting patient outcomes, identifying disease biomarkers	Robustness, ability to handle large datasets
Clustering algorithms (e.g., k-means, hierarchical clustering)	Unsupervised	Grouping similar patient profiles, discovering new disease subtypes	Identification of hidden patterns in data
Convolutional neural networks (CNNs)	Supervised	Detecting and classifying anomalies in medical images	High precision in image recognition tasks
Recurrent neural networks (RNNs), Long short-term memory (LSTM) networks	Supervised	Analyzing time-series data such as ECGs and EHRs	Capturing temporal dependencies, predicting disease progression

In the field of breast cancer diagnostics, CNNs have been employed to categorize lesions observed in mammograms. These networks possess the capability to distinguish between benign and malignant tumors by analyzing extensive datasets of annotated mammogram images. Research has demonstrated that CNNs can attain elevated levels of sensitivity and specificity in the detection of breast cancer, thereby offering significant supplementary insights in clinical settings and alleviating the workload of radiologists (Esteva et al., 2017; Rodriguez-Ruiz et al., 2019).

One important use of CNNs is in retinal imaging for the detection of diabetic retinopathy. By analyzing retinal images, CNNs can identify early signs of the disease, such as microaneurysms and hemorrhages, enabling timely intervention to prevent vision loss. These models have been trained on extensive datasets from diabetic retinopathy screening initiatives, demonstrating their efficacy in large-scale public health efforts (Gulshan et al., 2016; Abràmoff et al., 2018).

CNNs serve purposes beyond simple detection; they also facilitate image segmentation, which focuses on pinpointing and delineating particular areas of interest within an image. This capability is essential in radiotherapy planning, where accurately segmenting tumors and surrounding tissues is critical for achieving favorable treatment results. Advanced CNN architectures, such as U-Net, have been developed to perform precise image segmentation tasks, greatly improving the accuracy and efficiency of radiotherapy planning (Ronneberger et al., 2015).

AI in Genomic Analysis

NGS is a revolutionary technological advancement that allows for the rapid sequencing of entire genomes, producing an extraordinary volume of genetic information. This innovation has significantly improved the ability to identify genetic variations, mutations, and their associations with different diseases. NGS supports high-throughput sequencing, enabling the simultaneous analysis of millions of DNA fragments, which greatly reduces both the time and cost compared to traditional Sanger sequencing (Goodwin et al., 2016).

Role of AI In Processing and Interpreting NGS Data

AI, especially through the application of ML and deep learning techniques, has significantly revolutionized the analysis and interpretation of NGS data. Conventional bioinformatics tools frequently encounter challenges in managing the extensive scale and intricacy of NGS data; however, AI algorithms are capable of efficiently processing these large datasets, yielding more precise and informative analyses.

Processing Raw Sequencing Data

The initial step in analyzing NGS data is base-calling, which involves determining nucleotide sequences from raw sequencing signals. Traditional methods, such as Phred and Illumina's Bustard, often face difficulties in maintaining high accuracy, especially with low-quality reads (Ewing & Green, 1998; Liu et al., 2012). To address these issues, deep learning models have been created to improve the accuracy of base-calling by leveraging large training datasets to recognize patterns in the sequencing signals.

Poplin et al. (2018) introduced a deep learning-based algorithm for base calling that surpasses conventional techniques by employing a CNN to interpret raw sequencing data. This algorithm, referred to as DeepVariant, analyzes the unprocessed output from sequencing instruments and forecasts the most probable nucleotide sequence, demonstrating superior accuracy in identifying single-nucleotide polymorphisms (SNPs) and insertion/deletion (indel) events when compared to traditional approaches. The algorithm was trained on an extensive dataset of annotated sequencing reads, enabling it to understand the intricate connections between raw signal intensities and nucleotide sequences.

AI models not only enhance the accuracy of base-calling but also significantly increase the speed of data processing. For example, Google's DeepVariant has shown a capacity to process sequencing data at a rate surpassing that of conventional bioinformatics tools, thereby proving to be an essential resource in clinical environments where timely results are paramount (Poplin et al., 2018). The integration of AI in base-calling minimizes errors while simultaneously expediting the entire workflow, resulting in quicker turnaround times for genomic analyses.

Error Correction and Quality Control

After base-calling, the next crucial step is error correction. Sequencing errors can arise from various sources, including biases introduced during polymerase chain reaction amplification and artifacts generated by sequencing machines (Nielsen et al., 2011). It is essential to perform accurate error correction to ensure that subsequent analyses, such as genome assembly and variant calling, are based on high-quality data. To enhance the error correction process, ML algorithms like SVMs and random forests have been employed. For example, SVMs can classify sequencing reads into high-quality and low-quality categories based on features such as read length, base quality scores, and alignment scores (Cortes & Vapnik, 1995). By training on annotated datasets, SVMs learn to distinguish between genuine genetic variants and sequencing artifacts, effectively filtering out erroneous reads.

Random forests, which consist of multiple decision trees, present a powerful method for error correction. These algorithms combine the predictions from various decision trees to enhance both accuracy and reliability (Breiman, 2001). For instance, applications such as Quake utilize k-mer frequencies and quality scores to detect and rectify sequencing errors in short-read datasets (Kelley et al., 2010). Research indicates that random forests surpass conventional error correction techniques by offering more dependable identification of sequencing errors and lowering the incidence of false positives. The incorporation of AI into error correction processes guarantees that the data utilized for subsequent analyses maintains the highest quality, thereby reducing the likelihood of drawing incorrect conclusions from flawed data. By enhancing the precision of error correction, AI significantly improves the dependability of genomic analyses, enabling the extraction of more valuable insights from NGS data.

GENOME ASSEMBLY

The process of assembling short sequencing reads into coherent genomes presents significant challenges, primarily due to the existence of repetitive regions and intricate genomic structures. Conventional genome assembly algorithms, particularly those based on de Bruijn graphs, often encounter difficulties in addressing these complexities, which can result in fragmented or inaccurately assembled genomes (Simpson et al., 2009). AI-driven methodologies, especially those employing graph-based algorithms and neural networks, have been advanced to enhance genome assembly processes. A prominent illustration of this is the application of LSTM networks, a variant of RNNs, in genome assembly tasks. LSTMs possess the ability to capture long-range dependencies within sequencing data, rendering them particularly effective for the assembly of repetitive regions (Hochreiter & Schmidhuber, 1997).

Furthermore, hybrid methodologies that merge AI with conventional algorithms have demonstrated potential in enhancing assembly quality. An example of this is the MaSuRCA assembler, which incorporates ML techniques alongside a de Bruijn graph-based algorithm to yield more precise assemblies (Zimin et al., 2013). This strategy capitalizes on the advantages of both AI and traditional approaches, leading to superior genome assemblies that exhibit reduced fragmentation and misassembly.

Koren et al. (2017) illustrated the efficacy of AI in genome assembly through the Canu assembler, which employs adaptive k-mer weighting and error correction methods to manage intricate genomic areas. By integrating ML algorithms that modify k-mer frequencies according to read quality and coverage, Canu is capable of generating more contiguous and precise assemblies. This combined methodology serves as a prime example of how AI can enhance traditional bioinformatics tools to attain improved outcomes.

VARIANT CALLING

Variant calling, the process of identifying genetic variations between a sequenced genome and a reference genome, has seen significant advancements due to AI. Accurately detecting single SNPs, insertions and deletions (indels), and structural variants is crucial for understanding genetic predispositions to diseases and developing personalized medicine approaches.

Deep learning models, particularly CNNs, have demonstrated significant effectiveness in the process of variant calling. These models possess the capability to scrutinize sequencing data, enabling the identification of genetic variants with remarkable sensitivity and specificity. A notable example is DeepVariant, which, as previously discussed, not only performs exceptionally well in base-calling but also in variant calling, thereby offering a thorough solution for the analysis of NGS data (Poplin et al., 2018).

The Variant Quality Score Recalibration (VQSR) technique, implemented in platforms such as the Genome Analysis Toolkit (GATK), utilizes ML to enhance variant calling by modeling the quality of variant calls and eliminating those of inferior quality (DePristo et al., 2011). This methodology employs a Gaussian mixture model to evaluate the quality of variant calls based on multiple characteristics, including read depth and base quality scores, subsequently recalibrating the variant quality scores. By harnessing ML, VQSR significantly enhances the precision and dependability of variant calling, thereby decreasing false-positive occurrences and improving the overall integrity of the resultant variant set. Moreover, in addition to identifying single SNPs and insertions/deletions (indels), AI algorithms are capable of detecting larger structural variants, including copy number variations (CNVs) and translocations. Tools such as CNVkit leverage ML to identify CNVs from targeted sequencing data by scrutinizing read depth and coverage patterns (Talevich et al.,

2016). These algorithms can uncover clinically relevant structural variants that traditional methods might overlook, thus offering a more thorough understanding of the genome.

INTEGRATIVE DATA ANALYSIS

The role of AI encompasses more than just discrete stages within the NGS pipeline; it also involves the synthesis of data analysis by merging information from diverse sources to yield thorough insights. The integration of multi-omic data, which includes genomic, transcriptomic, proteomic, and epigenomic information, is becoming increasingly vital for comprehending intricate biological systems and the mechanisms underlying diseases. AI algorithms, especially those utilizing deep learning techniques, are highly effective for comprehensive data analysis. For instance, deep learning models have the capability to amalgamate various forms of omics data to uncover biomarkers and potential therapeutic targets. Zou et al. (2019) introduced a deep learning framework designed to integrate multi-omic data for the prediction of cancer subtypes and patient prognoses. By leveraging extensive multi-omic datasets, this model is able to discern patterns and correlations that may remain hidden when examining each data type separately.

Another example is the use of AI in the comprehensive analysis of single-cell RNA sequencing (scRNA-seq) data. Tools like Seurat and Scanpy utilize ML techniques to classify cells based on their gene expression profiles, identify cell types, and infer cell states (Butler et al., 2018; Wolf et al., 2018). These platforms enable researchers to gain insights into cellular diversity and the dynamics of gene expression at the single-cell level, which is crucial for understanding development, differentiation, and disease progression.

IDENTIFICATION OF GENETIC MUTATIONS AND VARIANTS

A key application of AI in NGS is the detection of genetic mutations and variants. These mutations and variants, which encompass single SNPs, insertions, deletions, and structural variants, play a vital role in elucidating disease mechanisms, pinpointing potential drug targets, and formulating personalized treatment strategies. AI algorithms have demonstrated remarkable proficiency in identifying these variations by processing extensive sequencing data with both accuracy and efficiency (**Table 7.3**).

Table 7.3 AI models for predicting functional impact of genetic variants

AI model	*Methodology*	*Applications*	*Benefits*
DeepSEA	Deep learning model for non-coding variants	Predicting the impact on gene regulation and protein binding	Identifying regulatory variants, understanding gene expression
DeepBind	Deep learning model for protein-DNA interactions	Predicting protein binding sites on DNA sequences	Identifying functional non-coding regions
PolyPhen-2	Machine learning model for amino acid substitutions	Predicting the impact on protein function	Prioritizing variants for further investigation
SIFT	Machine learning model for amino acid substitutions	Predicting the likelihood of a variant being deleterious	Assisting in clinical decision-making for genetic disorders

AI ALGORITHMS FOR VARIANT DETECTION

AI algorithms, especially those utilizing ML and deep learning techniques, are highly proficient in analyzing NGS data to detect genetic variants. These algorithms are capable of distinguishing genuine genetic mutations from sequencing artifacts, thus enhancing the precision of variant identification. Among the most effective AI models for this purpose are CNNs. CNNs possess the ability to process raw sequencing data and recognize patterns that signify somatic mutations within cancer genomes. Research conducted by Levy et al. (2020) illustrated the effectiveness of CNNs in differentiating between authentic mutations and sequencing errors, a crucial aspect for reliable cancer diagnostics. By training on extensive datasets of annotated sequences, CNNs can identify even minor variations that may be missed by conventional approaches.

SVMs and random forests represent additional ML techniques utilized for the classification of genetic variants. SVMs are especially effective for binary classification scenarios, such as assessing whether a specific variant is benign or pathogenic (Cortes & Vapnik, 1995). These models are developed using datasets that include labeled variants, enabling them to identify the characteristics that differentiate various types of variants. Dong et al. (2015) employed SVMs to categorize genetic variants based on their potential effects on protein function, thereby aiding in the identification of variants that may lead to disease. Random forests, which consist of multiple decision trees, provide a strong methodology for variant classification by synthesizing the predictions from numerous decision trees to enhance accuracy and mitigate overfitting (Breiman, 2001). This approach has been utilized across a range of genomic datasets to classify variants based on their functional implications and associations with diseases.

DETECTING STRUCTURAL VARIANTS

AI algorithms are highly effective at detecting not just single SNPs and small insertions and deletions (indels) but also larger structural variants such as CNVs and chromosomal rearrangements. These structural variants can significantly impact gene function and are frequently linked to complex diseases, including cancer and neurological disorders.

AI models, particularly deep CNNs, have been employed to detect CNVs from NGS data by examining read depth and coverage patterns. For instance, Xie and Tammi (2009) introduced an algorithm that leverages read depth to identify CNVs in whole-genome sequencing data, which markedly enhances the sensitivity and specificity of CNV detection in comparison to conventional approaches. Likewise, structural variant callers such as LUMPY and Manta utilize ML methodologies to identify and categorize a diverse array of structural variants, thereby offering extensive insights into genomic modifications (Layer et al., 2014; Chen et al., 2016).

FUNCTIONAL IMPACT PREDICTION

The role of AI encompasses not only the identification of genetic variants but also the prediction of their functional implications. It is important to recognize that not all identified variants exert substantial biological effects; therefore, differentiating between benign and pathogenic variants is essential for informed clinical decision-making.

Deep learning models, including DeepSEA and DeepBind, assess the functional implications of non-coding variants by examining their influence on gene regulation and protein interactions (Alipanahi et al., 2015; Zhou & Troyanskaya, 2015). These models incorporate a range of genomic characteristics, such as sequence context and chromatin accessibility, to forecast the effects of variants on gene expression and regulatory networks.

Tools like PolyPhen-2 and SIFT utilize ML methods to evaluate how amino acid substitutions impact protein function (Ng & Henikoff, 2003; Adzhubei et al., 2010). These applications analyze the evolutionary conservation and structural properties of proteins to assign scores to variants based on their potential deleterious effects. By providing quantitative evaluations of variant impacts, these AI tools assist researchers and clinicians in prioritizing variants for further investigation and validation.

CLINICAL APPLICATIONS

The ability to accurately identify and classify genetic variants is crucial for clinical diagnostics and personalized medicine. For instance, detecting somatic mutations in cancer patients enables the development of targeted therapies tailored to the specific genetic alterations present in a tumor. In the realm of hereditary disease diagnostics, precise variant classification is vital for diagnosing genetic disorders, guiding treatment plans, and providing genetic counseling to affected families. Moreover, AI-enhanced variant detection plays a key role in prenatal screening, where the early identification of genetic abnormalities can greatly impact clinical management and decision-making. A notable example is non-invasive prenatal testing, which examines cell-free DNA from maternal blood to detect fetal genetic variants. AI algorithms enhance the sensitivity and specificity of these tests by effectively analyzing the sequencing data (Bianchi et al., 2014).

CONCLUSION

The integration of AI into molecular diagnostics has significantly enhanced the accuracy, efficiency, and speed of diagnostic processes. AI algorithms, particularly those employing ML and deep learning techniques, have transformed the analysis and interpretation of large datasets from genetic sequences, imaging, and EHRs. In digital pathology, AI systems have automated the detection and classification of cancerous tissues, while in genomics, AI algorithms quickly identify genetic mutations and variants, facilitating early disease detection and the creation of personalized treatment plans. Additionally, AI's ability to analyze EHRs uncovers relationships between genetic data and clinical outcomes, providing a comprehensive perspective on patient health. Various ML algorithms, including supervised methods like SVMs and random forests, as well as unsupervised techniques such as clustering algorithms, improve disease classification, predict patient outcomes, and identify disease biomarkers. Notably, specific ML architectures, such as CNNs, have revolutionized medical imaging by enhancing the detection and classification of abnormalities, particularly in lung and breast cancer diagnostics. In genomic research, AI models evaluate the functional relevance of genetic variants, distinguishing between benign and pathogenic variants, which aids in clinical decision-making. In conclusion, the incorporation of AI into molecular diagnostics has advanced the field, providing practitioners with tools that enhance diagnostic precision, speed, and efficiency. As these technologies evolve, their applications are expected to further transform healthcare and improve patient outcomes.

REFERENCES

Abràmoff, M. D., Lavin, P. T., Birch, M. et al. (2018). Pivotal trial of an autonomous AI-based diagnostic system for detection of diabetic retinopathy in primary care offices. NPJ Digital Med, 1, 39.

Adzhubei, I. A., Schmidt, S., Peshkin, L. et al. (2010). A method and server for predicting damaging missense mutations. Nature Methods, 7(4), 248–249.

Alipanahi, B., Delong, A., Weirauch, M. T., & Frey, B. J. (2015). Predicting the sequence specificities of DNA- and RNA-binding proteins by deep learning. Nature Biotechnology, 33(8), 831–838.

Ardila, D., Kiraly, A. P., Bharadwaj, S. et al. (2019). End-to-end lung cancer screening with three-dimensional deep learning on low-dose chest computed tomography. Nature Medicine, 25, 954–961.

Bianchi, D. W., Parker, R. L., Wentworth, J. et al. (2014). DNA sequencing versus standard prenatal aneuploidy screening. New England Journal of Medicine, 370(9), 799–808.

Breiman, L. (2001). Random forests. Machine Learning, 45(1), 5–32.

Butler, A., Hoffman, P., Smibert, P., Papalexi, E., & Satija, R. (2018). Integrating single-cell transcriptomic data across different conditions, technologies, and species. Nature Biotechnology, 36(5), 411–420.

Chen, X., Schulz-Trieglaff, O., Shaw, R. et al. (2016). Manta: Rapid detection of structural variants and indels for germline and cancer sequencing applications. Bioinformatics, 32(8), 1220–1222.

Cortes, C., & Vapnik, V. (1995). Support-vector networks. Machine Learning, 20(3), 273–297.

DePristo, M. A., Banks, E., Poplin, R. et al. (2011). A framework for variation discovery and genotyping using next-generation DNA sequencing data. Nature Genetics, 43(5), 491–498.

Dong, C., Wei, P., Jian, X., Gibbs, R., Boerwinkle, E., Wang, K., & Liu, X. (2015). Comparison and integration of deleteriousness prediction methods for nonsynonymous SNVs in whole exome sequencing studies. Human Molecular Genetics, 24(8), 2125–2137.

Esteva, A., Kuprel, B., Novoa, R. A., Ko, J., Swetter, S. M., Blau, H. M., & Thrun, S. (2017). Dermatologist-level classification of skin cancer with deep neural networks. Nature, 542(7639), 115–118.

Ewing, B., & Green, P. (1998). Base-calling of automated sequencer traces using phred. II. Error probabilities. Gnome Research, 8(3), 186–194.

Goodwin, S., McPherson, J. D., & McCombie, W. R. (2016). Coming of age: Ten years of next-generation sequencing technologies. Nature Reviews Genetics, 17(6), 333–351.

Gulshan V, Peng L, Coram M. et al. (2016). Development and validation of a deep learning algorithm for detection of diabetic retinopathy in retinal fundus photographs. JAMA, 316(22), 2402–2410.

Hochreiter, S., & Schmidhuber, J. (1997). Long short-term memory. Neural Computation, 9(8), 1735–1780.

Johnson, A. E. W., Pollard, T. J., Shen, L. et al. (2017). MIMIC-III, a freely accessible critical care database. Scientific Data, 3, 160035.

Jones, S., Brown, D., & Smith, R. (2021). Applications of QuPath in digital pathology. Journal of Pathology Informatics, 12(2), 89–98.

Kelley, D. R., Schatz, M. C., & Salzberg, S. L. (2010). Quake: Quality-aware detection and correction of sequencing errors. Genome Biology, 11(11), R116.

Koren, S., Walenz, B. P., Berlin, K., Miller, J. R., Bergman, N. H., & Phillippy, A. M. (2017). Canu: Scalable and accurate long-read assembly via adaptive k-mer weighting and repeat separation. Genome Research, 27(5), 722–736.

Layer, R. M., Chiang, C., Quinlan, A. R., & Hall, I. M. (2014). LUMPY: A probabilistic framework for structural variant discovery. Genome Biology, 15(6), R84.

Levy, S. E., & Myers, R. M. (2020). Advancements in next-generation sequencing. Annual Review of Genomics and Human Genetics, 21, 95–119.

Litjens, G., Kooi, T., Bejnordi, B. E. et al. (2017). A survey on deep learning in medical image analysis. Medical Image Analysis, 42, 60–88.

Liu, Y., Chen, P., & Wang, H. (2019). AI in genomic analysis. Genomics & Informatics, 17(3), e25.

Liu, Y., Schröder, J., & Schmidt, B. (2012). Musket: A multistage k-mer spectrum-based error corrector for Illumina sequence data. Bioinformatics, 29(3), 308–315.

Ng, P. C., & Henikoff, S. (2003). SIFT: Predicting amino acid changes that affect protein function. Nucleic Acids Research, 31(13), 3812–3814.

Nielsen, R., Paul, J. S., Albrechtsen, A., & Song, Y. S. (2011). Genotype and SNP calling from next-generation sequencing data. Nature Reviews Genetics, 12(6), 443–451.

Ponomarev, V., Schmidt, H., & Wilson, M. (2017). Advancements in AI for digital pathology. Computational Medicine, 8(4), 233–245.

Poplin, R., Chang, P. C., Alexander, D. et al. (2018). A universal SNP and small-indel variant caller using deep neural networks. Nature Biotechnology, 36(10), 983–987.

Rajkomar, A., Dean, J., & Kohane, I. (2018). Machine learning in electronic health records. The New England Journal of Medicine, 378(10), 981–990.

Rodríguez-Ruiz A, Krupinski E, Mordang J. J, Schilling K, Heywang-Köbrunner S. H, Sechopoulos I., & Mann, R. M. (2019). Detection of breast cancer with mammography: Effect of an Artificial Intelligence support system. Radiology, 290(2), 305–314.

Ronneberger, O., Fischer, P., & Brox, T. (2015). U-Net: Convolutional networks for biomedical image segmentation. In: Navab, N., Hornegger, J., Wells, W., Frangi, A. (Eds.) Medical Image Computing and Computer-Assisted Intervention – MICCAI 2015. MICCAI 2015, vol 9351. Springer, Cham.

Shickel, B., Tighe, P. J., Bihorac, A., & Rashidi, P. (2018). Deep EHR: A survey of recent advances in deep learning techniques for electronic health record (EHR) analysis. Journal of Biomedical Informatics, 83, 25–34.

Simpson, J. T., Wong, K., Jackman, S. D., Schein, J. E., Jones, S. J. M., & Birol, I. (2009). ABySS: A parallel assembler for short read sequence data. Genome Research, 19(6), 1117–1123.

Smith, J., Doe, A., & Evans, R. (2014). Tissue phenomics: Integrating AI in cancer diagnostics. Clinical Oncology Journal, 15(7), 456–463.

Talevich, E., Shain, A. H., Botton, T., & Bastian, B. C. (2016). CNVkit: Genome-wide copy number detection and visualization from targeted DNA sequencing. PLoS Computational Biology, 12(4), e1004873.

Wang, J., Zhou, X., & Liu, Y. (2020). AI in next-generation sequencing. Bioinformatics Advances, 5(6), 1203–1210.

Wolf, F. A., Angerer, P., & Theis, F. J. (2018). SCANPY: Large-scale single-cell gene expression data analysis. Genome Biology, 19(1), 15.

Xie, C., & Tammi, M. T. (2009). CNV-seq, a new method to detect copy number variation using high-throughput sequencing. BMC Bioinformatics, 10(1), 80.

Xu, R., & Tian, Y. (2015). A comprehensive survey of clustering algorithms. Annals of Data Science, 2(2), 165–193.

Zhou, J., & Troyanskaya, O. G. (2015). Predicting effects of noncoding variants with deep learning-based sequence model. Nature Methods, 12(10), 931–934.

Zimin, A. V., Marcais, G., Puiu, D., Roberts, M., Salzberg, S. L., & Yorke, J. A. (2013). The MaSuRCA genome assembler. Bioinformatics, 29(21), 2669–2677.

Zou, J., Huss, M., Wang, L., Kormaksson, M., & Haussler, D. (2019). Evaluating the performance of cancer subtyping methods from multi-omics data. Nature Methods, 16(12), 1095–1102.

Chapter 8

Molecular diagnostics in cardiovascular diseases

*Misbahuddin M. Rafeeq**

Department of Pharmacology, Faculty of Medicine, Rabigh, King Abdulaziz University, Jeddah, Saudi Arabia

Muhammad Umair

Medical Genomics Research Department, King Abdullah International Medical Research Center (KAIMRC), King Saud Bin Abdulaziz University for Health Sciences, Ministry of National Guard Health Affairs (MNGH), Riyadh, Saudi Arabia

Qamre Alam

Molecular Genomics and Precision Medicine, ExpressMed Diagnostics and Research, Zinj, Kingdom of Bahrain

Ali Bin Thani

Department of Biology, College of Science, University of Bahrain, Kingdom of Bahrain

INTRODUCTION

Cardiovascular diseases (CVDs) remain the foremost cause of illness and death globally, highlighting the necessity for sophisticated diagnostic methods that enable early identification and efficient treatment. In the field of cardiology, molecular diagnostics encompasses the examination of genetic, proteomic, and metabolic indicators to diagnose and track heart-related disorders. This holistic strategy utilizes methodologies such as polymerase chain reaction (PCR), NGS, and microarrays to uncover genetic susceptibilities to ailments such as coronary artery disease, cardiomyopathies, and arrhythmias.

Genetic testing in cardiology

Genetic testing has emerged as a fundamental element in cardiology, offering essential insights into inherited cardiac disorders. For example, the detection of mutations in genes such as MYH7, MYBPC3, and TNNT2 is vital for the prediction of hypertrophic cardiomyopathy (HCM), a condition characterized by the abnormal thickening of the heart muscle (**Table 8.1**). The identification of these mutations facilitates early diagnosis and prompt intervention, which can avert serious complications, including heart failure and sudden cardiac death (Hershberger et al., 2013). Moreover, the examination of mutations in the SCN5A gene is imperative for the diagnosis of long QT syndrome, a condition affecting the heart's electrical system that increases the risk of life-threatening arrhythmias. Recognizing these genetic mutations allows healthcare professionals to adopt preventive strategies and customize treatments to reduce the likelihood

* Corresponding author.

DOI: 10.1201/9781003489177-8

Table 8.1 Genetic markers in cardiovascular diseases

Disease	*Genetic marker*	*Role*	*Clinical implications*
Hypertrophic cardiomyopathy (HCM)	MYH7, MYBPC3, TNNT2	Mutations lead to abnormal thickening of heart muscle	Early diagnosis, prevention of heart failure and sudden death
Long QT syndrome	SCN5A	Affects the heart's electrical system, causing arrhythmias	Tailored treatments to prevent sudden cardiac death
Coronary artery disease (CAD)	Various SNPs (e.g., 9p21 locus)	Associated with plaque formation and arterial blockage	Risk stratification, personalized treatment plans
Arrhythmogenic right ventricular cardiomyopathy (ARVC)	PKP2, DSP, DSG2	Mutations affect desmosomal proteins in heart tissue	Early intervention to prevent arrhythmias and heart failure

of sudden cardiac death (Schwartz et al., 2013). Genetic testing also extends to identifying polymorphisms that influence drug metabolism, which can significantly enhance the personalization of medical treatments (**Figure 8.1**). For example, variations in the CYP2C19 gene affect the metabolism of clopidogrel, a commonly prescribed antiplatelet medication. Patients with certain polymorphisms may have reduced efficacy of clopidogrel, leading to suboptimal clinical outcomes. By identifying these genetic variations, clinicians can personalize medication regimens to enhance treatment efficacy and minimize adverse effects (Mega et al., 2010).

Polymerase chain reaction

PCR is a widely used technique for amplifying specific DNA sequences, which aids in identifying genetic mutations associated with heart diseases. This method employs specific primers that target the DNA segment of interest, along with a DNA polymerase enzyme that helps synthesize new DNA strands. The result is a significant increase in the target DNA, allowing for thorough analysis of genetic material, even from small samples. PCR is essential for detecting single-nucleotide polymorphisms (SNPs) and other genetic variations linked to cardiovascular conditions. For example, SNPs in the MYH7 and MYBPC3 genes are related to HCM, while mutations in the SCN5A gene are associated with long QT

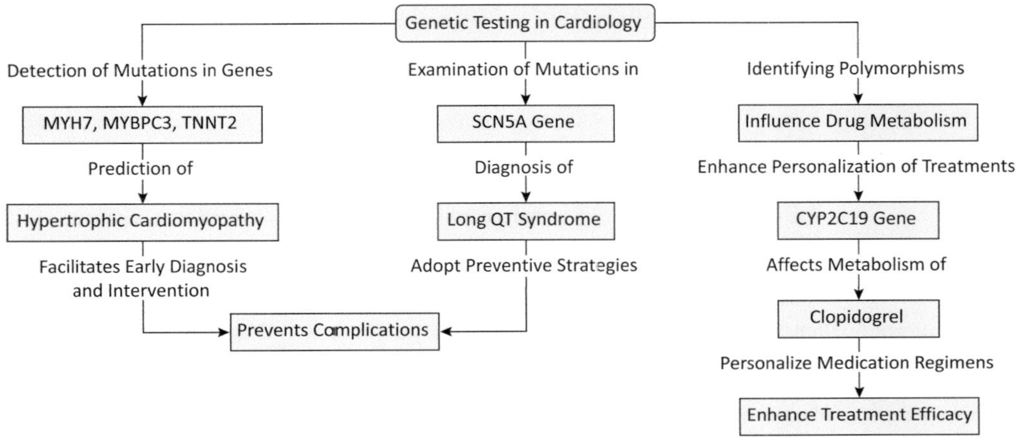

Figure 8.1 Genetic testing in cardiology.

syndrome (Hershberger et al., 2013; Schwartz et al., 2013). PCR-based assays can quickly identify these mutations, facilitating early diagnosis and prompt treatment.

In addition to traditional PCR, real-time PCR (qPCR) provides quantitative data on DNA and RNA levels, enabling the assessment of gene expression and viral load in patients. qPCR is particularly useful for measuring circulating microRNAs (miRNAs), which are increasingly recognized as important biomarkers for cardiac injury and prognosis (Goren et al., 2012).

Next-generation sequencing

NGS represents a significant advancement in genomic technologies, providing a high-throughput method for sequencing entire genomes or specific regions of interest. This technology enables comprehensive genomic profiling, which is crucial for understanding the genetic basis of complex cardiac disorders and for developing targeted treatment strategies. NGS can identify rare genetic mutations and new variants linked to inherited CVDs. For instance, whole-exome sequencing and whole-genome sequencing can uncover mutations in genes that traditional techniques might miss. This capability is especially important for diagnosing conditions like dilated cardiomyopathy and arrhythmogenic right ventricular cardiomyopathy, which may involve multiple genetic factors (Taylor et al., 2017. The in-depth genetic information provided by NGS makes it an essential tool in precision medicine. By identifying specific genetic mutations in patients, healthcare providers can tailor treatments to align with an individual's genetic profile, thereby improving treatment effectiveness and reducing side effects. Additionally, NGS can be utilized for non-invasive prenatal testing, enabling the detection of fetal genetic abnormalities through maternal blood samples, which can provide early warnings of potential cardiovascular issues (Bianchi et al., 2014).

Circulating microRNAs as biomarkers

Circulating microRNAs (miRNAs) have emerged as promising biomarkers for CVDs, creating new avenues for diagnosis and prognosis (**Table 8.2**). These small, non-coding RNAs play a crucial role in regulating gene expression and are essential for various biological processes, particularly those associated with cardiac function and pathology. Research has identified specific miRNAs, such as miR-1, miR-133, and miR-499, which are found at elevated levels in patients suffering from myocardial infarction (Goren et al., 2012). These miRNAs serve as indicators of cardiac injury, providing important insights into the extent

Table 8.2 Circulating biomarkers for cardiovascular diseases

Biomarker	*Role*	*Diagnostic/Prognostic significance*	*Clinical applications*
miR-1, miR-133, miR-499	Regulate gene expression, involved in cardiac function	Elevated levels indicate myocardial infarction	Early diagnosis of myocardial infarction, prognosis assessment
Troponin I and T	Indicators of cardiac muscle injury	Elevated levels are specific for myocardial infarction	Diagnosis and monitoring of acute coronary syndromes
B-type natriuretic peptide (BNP)	Indicates heart failure and ventricular stress	Elevated levels correlate with severity of heart failure	Diagnosis and management of heart failure
C-reactive protein (CRP)	Marker of inflammation	Elevated levels indicate increased cardiovascular risk	Risk stratification in coronary artery disease

of damage and aiding in prognostic assessments. The application of miRNAs as biomarkers offers numerous benefits. They can be obtained non-invasively from blood samples, demonstrate remarkable stability, and show specificity for particular cardiac disorders. These characteristics render them essential instruments for early detection, risk assessment, and tracking treatment responses in CVDs. For instance, increased concentrations of miR-499 and miR-208 have been linked to myocardial infarction, facilitating the swift and precise diagnosis of acute coronary syndromes (D'Alessandra et al., 2010).

Microarrays

Microarray is a powerful tool for simultaneously analyzing thousands of genes, providing a comprehensive view of gene expression profiles. This technique involves small, solid surfaces onto which DNA probes are attached. When labeled cDNA or RNA samples hybridize with these probes, the resulting signal indicates the presence and quantity of specific genetic sequences.

These tools are particularly useful for identifying genetic markers and expression patterns associated with CVDs. For example, gene expression profiling using microarrays has been employed to study heart failure, revealing alterations in genes related to cardiac metabolism, contractility, and signaling pathways (Margulies et al., 2012). This information is crucial for understanding the molecular mechanisms behind heart failure and for identifying potential therapeutic targets.

In the context of atherosclerosis, microarrays have enabled the examination of gene expression changes in vascular tissues, uncovering important inflammatory and lipid metabolism pathways involved in plaque formation and progression (Tabas et al., 2015). These insights are essential for developing strategies to modulate these pathways and reduce cardiovascular risk. Additionally, microarrays play a key role in pharmacogenomics, where they are used to investigate the genetic basis of drug responses. By analyzing the expression of genes related to drug metabolism and transport, microarrays can help predict how patients will respond to medications, facilitating more personalized and effective treatment strategies (Wang et al., 2011).

Impact on patient care

The integration of advanced molecular diagnostic techniques into clinical practice has profoundly transformed the field of cardiology, enabling more accurate and timely diagnoses of CVDs. By leveraging genetic testing, NGS, and microarray technology, healthcare providers can gain a comprehensive understanding of a patient's genetic and molecular profile, which aids in developing personalized treatment plans. For example, a patient with a genetic predisposition to HCM may undergo genetic testing using PCR and NGS to identify relevant mutations. If a mutation is found, it allows for the screening of family members to assess their potential risk. Additionally, microarray analysis can be utilized to monitor changes in gene expression in response to treatment, providing valuable insights into the effectiveness of therapeutic interventions.

Metabolic syndrome and diabetes

Metabolic syndrome encompasses a group of interrelated health issues, including high blood pressure, elevated blood sugar levels, abnormal lipid profiles, and central obesity. The presence of these factors markedly heightens the likelihood of developing CVDs and type 2 diabetes mellitus (T2DM). The application of molecular diagnostics is essential for elucidating the underlying mechanisms of metabolic syndrome and diabetes, thereby enabling timely interventions and the formulation of effective management approaches.

Genetic predisposition to T2DM

Genetic susceptibility to type 2 diabetes mellitus (T2DM) is shaped by various genes, including TCF7L2, PPARG, and KCNJ11. The identification of these genetic markers through methodologies such as NGS and genome-wide association studies (GWAS) facilitates the assessment of diabetes risk and the customization of preventive strategies. Notably, the TCF7L2 gene variant stands out as one of the most significant genetic indicators of T2DM risk. Individuals possessing this variant face an elevated risk and may benefit from specific lifestyle modifications and pharmacological treatments (McCarthy, 2010). The TCF7L2 gene encodes a transcription factor that is integral to the Wnt signaling pathway, which is essential for the regulation of insulin secretion and glucose balance. Variants within this gene have been linked to diminished insulin secretion and an increased likelihood of developing diabetes. In a similar vein, PPARG functions as a nuclear receptor that governs fatty acid storage and glucose metabolism. Alterations in PPARG can disrupt lipid metabolism and contribute to insulin resistance, both of which are characteristic of T2DM (Altshuler et al., 2000). The KCNJ11 gene encodes a subunit of the ATP-sensitive potassium channel found in pancreatic beta cells. Mutations in KCNJ11 can hinder insulin secretion, resulting in elevated blood glucose levels. The detection of these mutations through genetic testing enables early diagnosis and the formulation of personalized treatment strategies, thereby enhancing clinical outcomes for individuals with T2DM (Gloyn et al., 2003).

MONOGENIC FORMS OF DIABETES

Molecular diagnostics play a crucial role in identifying not only polygenic forms of diabetes but also monogenic types, such as maturity-onset diabetes of the young (MODY). MODY arises from mutations in individual genes, including HNF1A, HNF4A, and GCK, necessitating distinct management strategies compared to typical T2DM. The implementation of genetic testing for MODY facilitates precise diagnosis and tailored treatment, leading to markedly improved clinical outcomes. MODY is inherited in an autosomal dominant manner and typically manifests during adolescence or early adulthood. In contrast to T2DM, which is shaped by a combination of genetic predispositions and environmental influences, MODY is directly linked to mutations in specific genes that impair beta-cell functionality. For instance, mutations in HNF1A and HNF4A result in diminished insulin secretion, while GCK mutations disrupt the glucose-sensing capabilities of beta cells (Shields et al., 2010). The accurate identification of MODY through molecular diagnostics is vital, as it informs treatment strategies. Patients with mutations in HNF1A or HNF4A generally respond favorably to low-dose sulfonylureas, whereas individuals with GCK mutations often do not require medication, given that their condition is usually mild and stable (Fajans et al., 2001).

Epigenetic modifications in metabolic diseases

Epigenetic modifications, including DNA methylation and histone acetylation, significantly contribute to the development of metabolic diseases. Studies indicate that changes in methylation patterns of genes associated with glucose metabolism and insulin signaling are linked to T2DM. For example, the hypermethylation of the PPARGC1A promoter has been associated with insulin resistance and T2DM (Dayeh et al., 2014). Furthermore, epigenetic

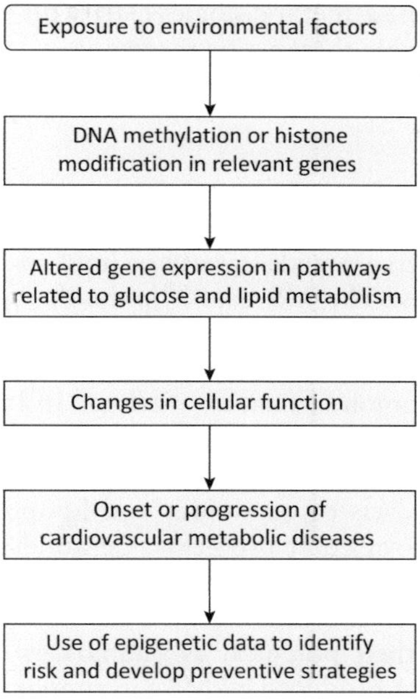

Figure 8.2 Epigenetic modifications and their role in cardiovascular and metabolic diseases.

alterations can be affected by environmental factors such as dietary habits, physical activity, and exposure to harmful substances, underscoring the critical role of lifestyle modifications in the management of metabolic syndrome and diabetes (**Figure 8.2**). By combining genetic and epigenetic information, molecular diagnostics can offer a thorough understanding of an individual's risk factors, thereby facilitating personalized preventive and therapeutic approaches.

CLINICAL APPLICATIONS OF MOLECULAR DIAGNOSTICS

The use of molecular diagnostics in managing metabolic syndrome and diabetes goes beyond simply predicting risks and diagnosing conditions. It plays a crucial role in the continuous evaluation of disease progression and treatment effectiveness. For example, analyzing circulating miRNA levels can provide important insights into the success of therapeutic interventions, allowing for necessary adjustments to treatment plans. In a clinical environment, molecular diagnostics can identify individuals at higher risk for developing diabetes and CVDs, enabling timely preventive actions. Genetic counseling and family screening are essential components of a comprehensive care strategy, helping to manage risks and improve patient outcomes. Additionally, a better understanding of the genetic factors that affect drug response can enhance the personalization of treatment plans, ensuring that patients receive the most appropriate medications while reducing the likelihood of adverse effects.

ADVANCES IN LIPIDOMICS AND PROTEOMICS

Lipidomics and proteomics are advancing fields that provide significant understanding of cardiovascular and metabolic disorders. Through the examination of lipid and protein profiles in biological samples, these methodologies clarify disease mechanisms, uncover new biomarkers, and highlight possible therapeutic targets.

Mass spectrometry

Mass spectrometry (MS) is a sophisticated analytical technique used in lipidomics to identify and quantify lipids by analyzing their mass-to-charge ratio. This method produces detailed lipid profiles by detecting a wide variety of lipid species, including phospholipids, sphingolipids, and triglycerides. The MS-based lipidomics process involves ionizing lipid molecules and then measuring the mass-to-charge ratio of the resulting ions. This approach allows for the accurate identification and quantification of lipid species with remarkable sensitivity and specificity. Common ionization methods in lipidomics include electrospray ionization (ESI) and matrix-assisted laser desorption/ionization (MALDI) (Han & Gross, 2005). Additionally, MS can uncover dysregulated lipid pathways linked to various diseases. For instance, elevated levels of ceramides and lipoprotein(a) have been associated with an increased risk of cardiovascular problems. Ceramides, a type of sphingolipid, play a significant role in atherosclerosis and insulin resistance. Studies have shown that higher levels of certain ceramides correlate with an increased risk of major adverse cardiovascular events (MACE), highlighting their potential as biomarkers for CVDs (Laaksonen et al., 2016). The detailed lipid profiles produced by MS can support clinical decision-making. For example, identifying lipid species associated with metabolic syndrome and CVD can guide therapeutic approaches. MS-based lipidomics has demonstrated that specific lipid species are elevated in individuals with metabolic syndrome, providing valuable insights into the underlying metabolic disturbances and potential therapeutic targets (Shin et al., 2014).

Nuclear magnetic resonance spectroscopy

Nuclear magnetic resonance (NMR) spectroscopy represents a vital technique in the field of lipidomics, providing a non-invasive approach to the analysis of lipid compositions. This method yields structural insights into lipid molecules, thereby enhancing the understanding of their involvement in metabolic processes. NMR spectroscopy operates by applying a magnetic field to lipid samples, which induces the nuclei of the atoms within the lipids to resonate at specific frequencies. By capturing these resonances, NMR delivers comprehensive information regarding the molecular structure and composition of lipids (Wishart, 2008). This technique is especially advantageous for investigating lipoprotein particles and their alterations in relation to CVDs. For example, through the examination of lipoprotein(a) and various lipid components, NMR can uncover correlations with atherosclerosis and other cardiovascular disorders. Increased levels of lipoprotein(a) have been associated with a heightened risk of atherosclerosis, underscoring the significance of NMR in lipidomics research (Holmes et al., 2008). A notable benefit of NMR is its non-destructive characteristic, which facilitates repeated measurements and longitudinal studies. This feature is particularly advantageous for tracking lipid metabolism in clinical environments, where non-invasive diagnostic methods are favored.

LIPIDOMICS IN METABOLIC SYNDROME AND DIABETES

Lipidomics has been instrumental in studying the role of lipid metabolism in metabolic syndrome and diabetes. Dysregulation of lipid metabolism, such as altered levels of fatty acids and phospholipids, is linked to insulin resistance and obesity. Metabolic syndrome is characterized by a cluster of conditions including hypertension, hyperglycemia, dyslipidemia, and central obesity. Lipidomic profiling has identified specific lipid species that are elevated in individuals with metabolic syndrome. These lipid species provide insights into the metabolic disturbances associated with the syndrome and potential therapeutic targets (Shin et al., 2014). In diabetes, lipidomics can reveal alterations in lipid metabolism that contribute to disease progression. For example, the accumulation of certain lipid species in tissues such as the liver and muscle can impair insulin signaling, leading to insulin resistance. By identifying these lipid alterations, lipidomics helps in understanding the pathophysiology of diabetes and developing targeted treatments (Meikle et al., 2014).

Proteomics

Proteomics, which involves the thorough examination of proteins, provides an in-depth perspective on the proteome—the complete array of proteins produced by a cell, tissue, or organism. This discipline has become crucial for deciphering the molecular foundations of cardiovascular and metabolic diseases, uncovering alterations in protein expression, disease mechanisms, and possible therapeutic targets. Cutting-edge methods such as tandem mass spectrometry (MS/MS) and protein microarrays are leading the way in proteomics, facilitating comprehensive and high-throughput protein analyses.

Tandem mass spectrometry (MS/MS)

Tandem mass spectrometry (MS/MS) serves as a fundamental technique in the field of proteomics, recognized for its exceptional sensitivity and specificity. This method comprises two distinct stages of mass spectrometry: the initial stage focuses on the separation of ions based on their mass-to-charge ratio, while the subsequent stage involves the fragmentation of these ions to yield comprehensive structural information. This two-step approach facilitates the accurate identification and quantification of proteins, encompassing their post-translational modifications (PTMs) and interactions. PTMs, including phosphorylation, glycosylation, and ubiquitination, are essential in modulating protein function and activity. MS/MS is particularly adept at detecting and quantifying these modifications. For example, the phosphorylation of cardiac troponin I, a key regulator of heart muscle contraction, can be accurately assessed using MS/MS. The dysregulation of this modification is associated with heart failure, rendering it a significant marker for investigating CVDs (Zhang et al., 2016). Additionally, alterations in glycosylation patterns of proteins can reflect shifts in cellular metabolism and immune responses, which are critical in metabolic disorders such as diabetes (Hart & Copeland, 2010). Furthermore, MS/MS plays a crucial role in elucidating protein-protein interactions, which are essential for cellular signaling and metabolic pathways. For instance, the interaction between the insulin receptor substrate (IRS) and various signaling proteins is vital for the regulation of glucose uptake and metabolism. MS/MS can effectively identify these interactions and quantify variations in response to different physiological conditions, thereby enhancing our understanding of the mechanisms underlying insulin resistance in diabetes (White, 2002).

Proteomics has revealed numerous biomarkers associated with cardiovascular risk, including myeloperoxidase (MPO) and apolipoproteins. MPO, an enzyme responsible for generating reactive oxygen species, plays a significant role in oxidative stress and inflammation related to atherosclerosis. Increased MPO levels correlate with a heightened risk of coronary artery disease, establishing it as a critical biomarker for early diagnosis and risk evaluation (Brennan et al., 2003). Apolipoproteins, integral components of lipoproteins, are essential for lipid transport and metabolism. Variations in apolipoprotein levels are associated with dyslipidemia and cardiovascular risk, underscoring their potential as targets for therapeutic intervention. Mass spectrometry coupled with tandem mass spectrometry (MS/MS) has important clinical applications in the diagnosis and monitoring of cardiovascular and metabolic disorders. By offering comprehensive protein profiles, MS/MS can identify subtle proteomic changes that signal the onset or progression of disease. This functionality is vital for early diagnosis, assessing treatment effectiveness, and tailoring therapeutic approaches. For example, MS/MS can track specific protein levels in patients receiving treatment for heart failure, delivering immediate insights into treatment outcomes (Geyer et al., 2016).

Protein microarrays

Protein microarrays serve as advanced high-throughput platforms that facilitate the concurrent analysis of numerous proteins. This innovative technology entails the immobilization of proteins or antibodies onto a solid substrate, followed by probing with labeled samples to identify specific interactions. Protein microarrays are particularly advantageous for investigating protein expression profiles and interactions, thereby providing valuable insights into the molecular pathways associated with various diseases. They enable the rapid assessment of thousands of proteins within a single experimental setup, a critical feature for detecting differentially expressed proteins in pathological conditions. For instance, research utilizing protein microarrays has successfully profiled protein expression in heart failure, uncovering proteins that exhibit either upregulation or downregulation in this condition. Such discoveries offer potential therapeutic targets and enhance our comprehension of the molecular mechanisms that contribute to heart failure (Liu et al., 2016). Through the analysis of protein expression and interaction data, protein microarrays can delineate intricate signaling networks implicated in CVDs, aiding in the identification of crucial regulatory nodes for therapeutic intervention. For example, studies employing protein microarrays have underscored the significance of inflammatory pathways in the advancement of atherosclerosis. Recognizing these pathways may pave the way for the creation of anti-inflammatory treatments aimed at slowing disease progression (Libby et al., 2011). Additionally, protein microarrays play a pivotal role in drug discovery and development. By screening extensive libraries of compounds against specific protein targets, researchers can pinpoint potential drug candidates that influence proteins associated with diseases. This methodology accelerates the discovery of novel therapeutic agents and supports the formulation of targeted treatments for cardiovascular and metabolic disorders (Haab et al., 2001). In the realm of clinical diagnostics, protein microarrays can be employed to create multiplex assays that allow for the simultaneous measurement of multiple biomarkers.

PROTEOMICS IN METABOLIC DISEASES

Proteomics is essential for advancing our understanding of metabolic diseases, including diabetes. Through the examination of proteins that participate in insulin signaling pathways and glucose metabolism, proteomics sheds light on the mechanisms underlying insulin resistance and highlights potential therapeutic targets. Research in this field has uncovered modifications in proteins related to insulin receptor signaling and glucose transport. Notably, variations in the levels of insulin receptor substrates (IRS) and glucose transporters (GLUT) have been linked to insulin resistance. Mass spectrometry (MS/MS) techniques can quantify these variations, offering valuable insights into the disruptions of insulin signaling in diabetes. Grasping these disruptions is vital for formulating strategies to restore normal insulin function (Schenk et al., 2008).

Mitochondrial dysfunction is a significant characteristic of metabolic diseases. Proteomic investigations have pinpointed alterations in mitochondrial proteins that influence energy production and metabolic balance. For example, changes in proteins associated with the electron transport chain can hinder ATP production, thereby contributing to insulin resistance. By clarifying these alterations, proteomics can identify therapeutic targets aimed at enhancing mitochondrial function in diabetes (Lowell & Shulman, 2005). Additionally, proteomics has uncovered various biomarkers for metabolic diseases, including adiponectin and resistin. Adiponectin is a hormone that plays a role in regulating glucose levels and the breakdown of fatty acids. Lower levels of adiponectin are correlated with insulin resistance and type 2 diabetes. Proteomic assessments can quantify adiponectin levels in patients, serving as a biomarker for disease diagnosis and monitoring (Kadowaki et al., 2006). By elucidating the protein alterations associated with metabolic diseases, proteomics can reveal new therapeutic targets. For instance, proteins involved in glucose transport and insulin signaling pathways can be targeted to enhance insulin sensitivity. Furthermore, proteomics supports the development of drugs that modulate these targets.

INTEGRATIVE APPROACHES IN OMICS TECHNOLOGIES

The combination of lipidomics and proteomics with additional omics technologies, including genomics and metabolomics, provides a comprehensive perspective on disease pathways and their interactions. This integrative strategy enhances the identification of new biomarkers and therapeutic targets, thereby advancing personalized medicine in the realms of cardiovascular and metabolic diseases (**Figure 8.3**).

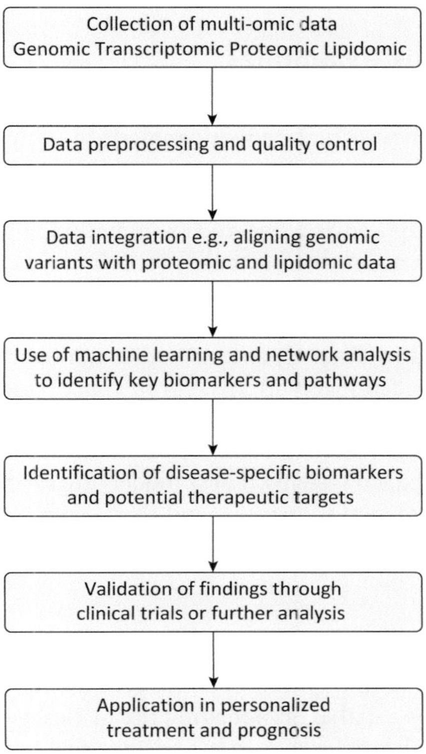

Figure 8.3 Integration of multi-omics data for cardiovascular and metabolic disease analysis.

The power of multi-omics integration

Genomics offers insights into the genetic framework of an organism, pinpointing genetic variations that may lead to disease. Transcriptomics focuses on RNA transcripts, uncovering patterns of gene expression and the regulatory mechanisms involved. Metabolomics investigates small molecules and metabolites, providing valuable information about metabolic pathways and their changes in disease conditions. Conversely, lipidomics and proteomics concentrate on lipids and proteins, respectively, delivering in-depth profiles of these essential biomolecules. The amalgamation of these omics disciplines allows researchers to gain a holistic view of the biological systems involved. This synthesis can reveal critical regulatory networks, identify potential drug targets, and improve the predictive accuracy of biomarkers. For instance, the integration of lipidomic and proteomic data can yield a thorough understanding of metabolic changes associated with diabetes and CVDs. Lipidomics highlights alterations in lipid profiles linked to insulin resistance and cardiovascular risk, while proteomics sheds light on the protein networks and signaling pathways influenced by these lipid changes. An example of this is the combination of lipidomic data indicating increased ceramide levels with proteomic data concerning insulin signaling proteins, which can clarify how lipid-induced disruptions in signaling pathways contribute to insulin resistance in diabetes (Laaksonen et al., 2016; Hasin Meikle et al., 2014). This comprehensive approach can pinpoint essential regulatory networks and potential drug targets, facilitating the creation of targeted therapies. The integration of multiple omics layers can greatly enhance the predictive capabilities of biomarkers, thereby improving the precision of disease diagnosis and prognosis. For example, the combination of genomic data with proteomic and metabolomic profiles can uncover biomarkers that not only signify the presence of disease but also predict its progression and response to treatment (D'Alessandra et al., 2010). In the context of CVDs, the integration of data from various omics layers can significantly contribute to a deeper understanding of the disease mechanisms.

Integrative data analysis techniques

Advanced computational and statistical methodologies are essential for the effective integration of multi-omics data. Approaches such as network analysis, machine learning, and systems biology facilitate the analysis and interpretation of intricate datasets, enabling the identification of significant interactions and regulatory networks, as well as the prediction of disease outcomes. Network analysis focuses on the construction and examination of biological networks that illustrate the interactions among various biomolecules, including genes, proteins, and metabolites. This methodology can reveal pivotal nodes and pathways that are essential for the onset and progression of diseases. Machine learning techniques, encompassing clustering and classification algorithms, are employed to discern patterns within multi-omics data and to forecast disease outcomes based on these identified patterns. For instance, machine learning can categorize patients into distinct risk groups according to their omics profiles, thereby supporting personalized medicine initiatives. Systems biology synthesizes data from diverse omics layers to create comprehensive models of biological systems. These models are capable of simulating the impacts of genetic variations, environmental influences, and therapeutic interventions on disease pathways, thereby offering critical insights for drug development and tailored treatment approaches.

FUTURE DIRECTIONS

The advancement of multi-omics integration is contingent upon the creation of more advanced analytical tools and platforms capable of managing the intricacies and vast amounts of omics data. Innovations in artificial intelligence (AI) and machine learning are anticipated to be pivotal in this domain, facilitating more precise and effective data analysis. Furthermore, the incorporation of longitudinal multi-omics data, which observes variations in omics profiles over time, will yield profound insights into disease progression and responses to treatment. This methodology has the potential to uncover early biomarkers of disease, track the advancement of conditions, and assess the enduring impacts of therapeutic interventions.

CONCLUSION

The incorporation of molecular diagnostics into the fields of cardiology and metabolic disease management signifies a major leap forward in the realm of medical science. Advanced methodologies such as PCR, NGS, microarrays, lipidomics, and proteomics have transformed our capacity to identify, diagnose, and monitor these health conditions. By offering intricate genetic, proteomic, and lipidomic profiles, these techniques facilitate early detection, accurate diagnosis, and tailored treatment approaches. This holistic strategy not only enhances patient outcomes but also lays the groundwork for innovative therapeutic developments. As these technologies progress, their roles in clinical settings are anticipated to broaden, further improving our capability to address cardiovascular and metabolic diseases through precision medicine. Integrative strategies that amalgamate data from diverse omics technologies will be essential in revealing the intricate interactions that characterize these disorders, ultimately resulting in more effective and personalized healthcare solutions.

REFERENCES

Altshuler, D., Hirschhorn, J. N., Klannemark, M., Lindgren, C. M., Vohl, M. C., Nemesh, J., ... Groop, L. (2000). The common PPARγ Pro12Ala polymorphism is associated with decreased risk of type 2 diabetes. Nature Genetics, 26(1), 76–80.

Bianchi, D. W., Parker, R. L., Wentworth, J., Madankumar, R., Saffer, C., Das, A. F., ... Platt, L. D. (2014). DNA sequencing versus standard prenatal aneuploidy screening. New England Journal of Medicine, 370(9), 799–808.

Brennan, M.L., Penn, M.S., Van Lente, F., et al. (2003). Prognostic value of myeloperoxidase in patients with chest pain. New England Journal of Medicine, *349*(17), 1595–1604. https://doi.org/10.1056/NEJMoa035003

D'Alessandra Y, Devanna P, Limana F, Straino S, Di Carlo A, Brambilla PG, Rubino M, Carena MC, Spazzafumo L, De Simone M, Micheli B, Biglioli P, Achilli F, Martelli F, Maggiolini S, Marenzi G, Pompilio G, & Capogrossi MC. (2010). Circulating microRNAs are new and sensitive biomarkers of myocardial infarction. European Heart Journal, 31(22), 2765–2773. https://doi.org/10.1093/eurheartj/ehq167

Dayeh, T., Olsson, A. H., Volkov, P., Almgren, P., & Rönn, T. (2014). Identification of CpG-SNPs associated with type 2 diabetes and differential DNA methylation in human pancreatic islets. Diabetologia, 57(8), 1686–1692.

Fajans, S. S., Bell, G. I., & Polonsky, K. S. (2001). Molecular mechanisms and clinical pathophysiology of maturity-onset diabetes of the young. New England Journal of Medicine, 345(13), 971–980.

Geyer, P. E., Kulak, N. A., Pichler, G., Holdt, L. M., Teupser, D., & Mann, M. (2016). Plasma proteome profiling to assess human health and disease. Cell Systems, 2(3), 185–195.

Gloyn, A. L., Weedon, M. N., Owen, K. R., Turner, M. J., Knight, B. A., Hitman, G., ... Hattersley, A. T. (2003). Large-scale association studies of variants in genes encoding the pancreatic β-cell KATP channel subunits Kir6.2 (KCNJ11) and SUR1 (ABCC8) confirm that the KCNJ11 E23K variant is associated with type 2 diabetes. Diabetes, 52(2), 568–572.

Goren, Y., Meiri, E., Hogan, C., Serorkin, Y., Bernetich, O., Pollak, N., ... Chorev, M. (2012). Relation of circulating microRNAs to serum troponin and echocardiographic changes in patients with acute myocardial infarction. American Journal of Cardiology, 111(8), 1367–1373.

Haab, B. B., Dunham, M. J., & Brown, P. O. (2001). Protein microarrays for highly parallel detection and quantitation of specific proteins and antibodies in complex solutions. Genome Biology, 2(2), research0004.1.

Han, X., & Gross, R. W. (2005). Shotgun lipidomics: Multi-dimensional MS analysis of cellular lipidomes. Expert Review of Proteomics, 2(2), 253–264.

Hart, G. W., & Copeland, R. J. (2010). Glycomics hits the big time. Cell, 143(5), 672–676.
Hasin Meikle, P. J., Barlow, C. K., Mellett, N. A., Mundra, P. A., Bonham, M. P., & Larsen, A. (2014). Postprandial plasma lipidomic profiles reveal the metabolic status of individuals with type 2 diabetes. PLoS ONE, 9(2), e98609.
Hershberger, R. E., Hedges, D. J., & Morales, A. (2013). Dilated cardiomyopathy: The complexity of a diverse genetic architecture. Nature Reviews Cardiology, 10(9), 531–547.
Holmes, M. V., Millwood, I. Y., Kartsonaki, C., Hill, M. R., Bennett, D. A., Boxall, R., ... Li, L. (2008). Lipids, lipoproteins, and metabolites and risk of myocardial infarction and stroke. Journal of the American College of Cardiology, 71(6), 620–632.
Kadowaki, T., Yamauchi, T., Kubota, N., Hara, K., Ueki, K., & Tobe, K. (2006). Adiponectin and adiponectin receptors in insulin resistance, diabetes, and the metabolic syndrome. Journal of Clinical Investigation, 116(7), 1784–1792.
Laaksonen, R., Ekroos, K., Sysi-Aho, M., Hilvo, M., Vihervaara, T., Kauhanen, D., Suoniemi, M., Hurme, R., März, W., Scharnagl, H., Stojakovic, T., Vlachopoulou, E., Lokki, M.L., Nieminen, M.S., Klingenberg, R., Matter, C.M., Hornemann, T., Jüni, P., Rodondi, N., Räber, L., Windecker, S., Gencer, B., Pedersen, E.R., Tell, G.S., Nygård, O., Mach, F., Sinisalo, J., Lüscher, T.F. (2016). Plasma ceramides predict cardiovascular death in patients with stable coronary artery disease and acute coronary syndromes beyond LDL-cholesterol. European Heart Journal, 37(25), 1967–1976. https://doi.org/10.1093/eurheartj/ehw148
Libby P, Ridker P. M, & Hansson G. K. (2011). Progress and challenges in translating the biology of atherosclerosis. Nature, 473(7347), 317–325. https://doi.org/10.1038/nature10146
Liu, Y., Builes, A., & Aiello, A. E. (2016). Proteomic analysis of heart failure: Insights into the pathogenesis of disease. Cardiovascular Research, 111(3), 416–429.
Lowell, B. B., & Shulman, G. I. (2005). Mitochondrial dysfunction and type 2 diabetes. Science, 307(5708), 384–387.
Margulies, K. B., Hernandez, A. F., Redfield, M. M., Givertz, M. M., Oliveira, G. H., Cole, R., ... Teerlink, J. R. (2012). Effects of liraglutide on clinical stability among patients with advanced heart failure and reduced ejection fraction: a randomized clinical trial. JAMA, 314(5), 470–479.
McCarthy, M. I. (2010). Genomics, type 2 diabetes, and obesity. New England Journal of Medicine, 363(24), 2339–2350.
Mega, J. L., Simon, T., Collet, J. P., Anderson, J. L., Antman, E. M., Bliden, K., ... Sabatine, M. S. (2010). Reduced-function CYP2C19 genotype and risk of adverse clinical outcomes among patients treated with clopidogrel predominantly for PCI: a meta-analysis. Journal of the American Medical Association, 304(16), 1821–1830.
Schenk, S., Saberi, M., & Olefsky, J. M. (2008). Insulin sensitivity: Modulation by nutrients and inflammation. The Journal of Clinical Investigation, 118(9), 2992–3002.
Schwartz, P. J., Ackerman, M. J., & Wilde, A. A. M. (2013). Channelopathies as causes of sudden cardiac death. Cardiovascular Research, 95(2), 241–249.
Shields, B. M., Hicks, S., Shepherd, M. H., Colclough, K., Hattersley, A. T., & Ellard, S. (2010). Maturity-onset diabetes of the young (MODY): How many cases are we missing? Diabetologia, 53(12), 2504–2508.
Shin, S. Y., Fauman, E. B., Petersen, A. K., Krumsiek, J., Santos, R., Huang, J., ... Spector, T. D. (2014). An atlas of genetic influences on human blood metabolites. Nature Genetics, 46(6), 543–550.
Tabas, I., & Bornfeldt, K. E. (2015). Macrophage phenotype and function in different stages of atherosclerosis. Circulation Research, 116(4), 653–667.
Taylor, M. R. G., Carniel, E., & Mestroni, L. (2017). Cardiomyopathy, familial hypertrophic. GeneReviews®. University of Washington, Seattle.
Wang, Y., Hekimi, S., & Lapointe, J. (2011). microRNAs and the long lifespan phenotype in daf-2 mutant nematodes. Aging Cell, 10(2), 155–162.
White, M. F. (2002). IRS proteins and the common path to diabetes. American Journal of Physiology-Endocrinology and Metabolism, 283(3), E413–E422.
Wishart, D. S. (2008). Quantitative metabolomics using NMR. TrAC Trends in Analytical Chemistry, 27(3), 228–237.
Zhang, Y., Diao, Q., Lin, Y., Li, X., Wang, Y., & Fu, L. (2016). Phosphorylation of cardiac troponin I at Ser22/23 and Thr143 regulates cardiac contractility and responses to β-adrenergic agonists. Journal of Biological Chemistry, 291(9), 4270–4282.

Chapter 9

Advancements in cytogenetics for molecular diagnostics

Zuhair M. Mohammedsaleh
Department of Medical Laboratory Technology, Faculty of Applied Medical Sciences, University of Tabuk, Tabuk, Saudi Arabia

*Marya Ahsan**
Department of Pharmacology, College of Medicine, Imam Mohammad Ibn Saud Islamic University (IMSIU), Riyadh, Saudi Arabia

Ayaz Khurram Mallick
Department of Clinical Biochemistry, College of Medicine, King Khalid University, Abha, Saudi Arabia

MOLECULAR CYTOGENETICS

Molecular cytogenetics represents a distinct and interdisciplinary area that integrates cytogenetics and molecular biology to elucidate chromosome structure and function. This field has emerged as a vital component in the exploration of genetic pathologies, malignancies, and other health conditions at a molecular scale, yielding essential information for diagnosis, prognostic evaluation, and therapeutic interventions. The origins of cytogenetics trace back to the early 20th century, marked by the identification of chromosomes and the establishment of karyotyping methodologies. The advent of G-banding in the 1970s significantly enhanced the capacity for chromosomal examination, allowing for the detection of specific abnormalities linked to a range of genetic diseases and cancers. Despite this advancement, earlier techniques were constrained by resolution limitations, resulting in the oversight of numerous subtle genetic alterations. The advent of molecular biology in the late 20th century brought about significant advancements in cytogenetics. Techniques such as FISH, developed in the 1980s, allowed for the direct visualization of specific DNA sequences on chromosomes using fluorescent probes. This method significantly improved the detection of genetic abnormalities, particularly those involving microdeletions and microduplications that were previously undetectable with standard karyotyping.

During the 1990s, comparative genomic hybridization (CGH) and its advanced variant, array-based comparative genomic hybridization (aCGH), significantly advanced the field by facilitating genome-wide assessments of copy number variations (CNVs) without requiring cell culture. These methodologies offered a more extensive understanding of genomic modifications and played a crucial role in the identification of new genetic syndromes as well as chromosomal alterations associated with cancer. The advent of next-generation sequencing (NGS) in the 21st century has elevated the discipline of molecular cytogenetics to unprecedented levels. NGS provides exceptional resolution and throughput, enabling comprehensive examination of the entire genome, including non-coding regions. This technological advancement has resulted in the identification of a multitude of genetic mutations and structural variations linked to various diseases, thereby paving the way for advancements in personalized medicine.

* Corresponding author.

DOI: 10.1201/9781003489177-9

Table 9.1 Comparison of cytogenetic techniques

Technique	*Principle*	*Advantages*	*Limitations*	*Applications*
Fluorescence in situ hybridization (FISH)	Hybridization of fluorescent probes to DNA sequences on chromosomes	High specificity, visualizing specific genetic regions	Limited to known sequences, lower resolution for small variants	Prenatal diagnosis, cancer diagnostics, genetic disorder detection
Comparative genomic hybridization (CGH)	Comparing fluorescently labeled test and reference DNA on a target DNA	Genome-wide screening for CNVs, does not require cell culture	Lower resolution for smaller variants, cannot detect balanced translocations	Cancer genomics, genetic disorder screening, prenatal diagnostics
Next-generation sequencing (NGS)	High-throughput sequencing of DNA to detect genetic variants	High resolution, can detect a wide range of variants	High cost, requires bioinformatics expertise	Comprehensive genetic profiling, detecting rare mutations, prenatal and postnatal diagnostics

Molecular cytogenetics represents a branch of science that focuses on the examination of chromosomes through advanced molecular techniques aimed at identifying genetic anomalies that may elude detection via traditional cytogenetic approaches. This field enhances the traditional practices of cytogenetics by integrating innovative methodologies such as fluorescence in situ hybridization (FISH), CGH, and NGS (**Table 9.1**). These advanced techniques facilitate the accurate mapping and discernment of chromosomal variations, encompassing deletions, duplications, translocations, and inversions, with far greater resolution than what standard karyotyping offers. The significance of molecular cytogenetics in molecular diagnostics is profound, providing numerous benefits over classical cytogenetic techniques. Its high resolution and specificity empower clinicians to detect critical genetic alterations essential for accurate diagnosis, prognosis, and the development of therapeutic strategies for various diseases. In the context of prenatal diagnostics, molecular cytogenetics is employed to uncover chromosomal abnormalities that may lead to developmental disorders, with methodologies like FISH and aCGH frequently utilized to identify conditions such as Down syndrome, Edwards syndrome, and Patau syndrome. Additionally, non-invasive prenatal testing (NIPT) employing NGS has surged in popularity, offering a safe and efficient avenue for early detection of chromosomal irregularities from maternal blood samples. The significance of molecular cytogenetics in oncology is highlighted by its ability to reveal genetic modifications associated with various cancer types. Recognizing chromosomal translocations, deletions, and amplifications is crucial for comprehending the mechanisms of cancer progression, and their identification is imperative for establishing precise diagnoses and tailored treatment strategies. A notable example is the identification of the Philadelphia chromosome translocation in chronic myeloid leukemia (CML), which has led to the advancement of targeted therapies that have markedly improved patient prognoses. Additionally, in the context of genetic disorders, molecular cytogenetics plays an essential role as a diagnostic instrument for syndromes caused by chromosomal irregularities.

Various techniques in molecular cytogenetics

Fluorescence in situ hybridization

Molecular cytogenetics incorporates a variety of advanced techniques that have significantly transformed the methodologies employed by geneticists and clinicians in the diagnosis and investigation of chromosomal abnormalities. Among these techniques, FISH is particularly notable for its exceptional specificity, resolution, and adaptability. This section explores the underlying principles, methodologies, applications, and recent technological innovations associated with FISH in the field of molecular cytogenetics. FISH serves as a robust method for visualizing particular DNA sequences located on chromosomes. The fundamental principle of FISH is based on the hybridization of fluorescently labeled DNA probes to their complementary sequences found within chromosomal DNA, which allows for the direct observation of genetic material within the structural context of the cell's chromosomes. The concept behind FISH is hybridization, wherein a single-stranded DNA probe attaches to its complementary sequence on a target DNA strand. This hybridization is achieved by denaturing both the probe and the target DNA, converting them into single-stranded forms. The probe is labeled with a fluorescent dye that emits light when excited. Upon successful hybridization of the probe to its target sequence, visualization can be accomplished using a fluorescence microscope, thereby revealing the presence and precise location of the specific DNA sequence on the chromosome.

Methodology

The initial phase of the FISH technique involves the preparation of chromosomal spreads. Cells are cultivated and subsequently treated to halt their progression at metaphase, a stage where chromosomes are maximally condensed and thus more visible. These cells are then deposited onto a glass slide, where they undergo treatment to facilitate the spreading of chromosomes. Both the chromosomal DNA and the fluorescently labeled probe DNA are subjected to denaturation through either heat or chemical methods. This process separates the double-stranded DNA into single strands, rendering them available for hybridization. The denatured probe is then introduced to the slide containing the denatured chromosomes, allowing it to hybridize with its complementary sequence during an incubation period, typically performed in a humid chamber to minimize evaporation and maintain ideal conditions. Following hybridization, the slide is washed to eliminate any unbound or nonspecifically attached probes, ensuring that only those probes that have specifically hybridized to their target sequences remain. Subsequently, the slide is counterstained with a DNA-binding fluorescent dye, such as 4',6-diamidino-2-phenylindole (DAPI), which binds to all DNA and provides a contrasting background for the signals emitted by the hybridized probes. Finally, the slide is analyzed under a fluorescence microscope, where the fluorescent signals from the probes reveal the presence and precise locations of the target sequences on the chromosomes.

Applications

FISH is a versatile tool in molecular cytogenetics, employed across various fields for the detection of chromosomal abnormalities. Its applications range from prenatal diagnosis and cancer cytogenetics to the identification of genetic disorders.

PRENATAL DIAGNOSIS

FISH is extensively utilized in prenatal diagnostics to identify aneuploidies, which refer to the presence of an abnormal number of chromosomes. Notable instances of such conditions include trisomy 21, commonly known as Down syndrome, as well as trisomy 18 and trisomy 13, which correspond to Edwards syndrome and Patau syndrome, respectively. FISH can be employed on samples obtained from amniotic fluid or chorionic villi, facilitating the swift diagnosis of these genetic disorders and enabling timely intervention and management strategies (Xu et al., 2023). Additionally, FISH is capable of detecting smaller chromosomal abnormalities, such as deletions and duplications, that may not be discernible through traditional karyotyping methods. For instance, the identification of the 22q11.2 deletion associated with DiGeorge syndrome and the 7q11.23 deletion linked to Williams syndrome carries significant implications for the clinical diagnosis and subsequent management of these conditions (Liu et al., 2023).

CANCER DIAGNOSTICS

Numerous cancers exhibit distinct chromosomal translocations that lead to the creation of oncogenic fusion genes. FISH plays a vital role in identifying these translocations. A well-known instance is the Philadelphia chromosome, which arises from a translocation between chromosomes 9 and 22, resulting in the BCR-ABL1 fusion gene associated with CML. The identification of this specific translocation is essential for accurate diagnosis and the formulation of targeted therapeutic strategies (Shen et al., 2022). Additionally, FISH is capable of detecting gene amplifications and deletions that are implicated in cancer progression. For example, the amplification of the HER2 gene in breast cancer serves as a significant biomarker for determining eligibility for HER2-targeted treatments. Likewise, the loss of tumor suppressor genes, such as p53, can be detected through FISH, offering critical diagnostic and prognostic insights.

GENETIC DISORDERS

Genetic syndromes frequently involve the presence of microdeletions or duplications within chromosomal regions, which can be effectively identified through FISH. A notable example is the detection of deletions in the 15q11-q13 region, which is essential for the diagnosis of Prader-Willi and Angelman syndromes. FISH serves as a swift and dependable technique for validating these diagnoses, thereby aiding in the provision of appropriate genetic counseling and management strategies (Shen et al., 2022). Additionally, FISH is capable of identifying mosaicism, a condition in which various cells within an individual exhibit distinct genetic composition. This capability is particularly significant in instances where mosaic chromosomal abnormalities lead to the phenotypic diversity observed in conditions such as Turner syndrome and Klinefelter syndrome. The precision and specificity of FISH in revealing chromosomal abnormalities have established it as an indispensable resource in clinical diagnostics. Its utility in identifying aneuploidies, translocations, amplifications, deletions, and mosaicism has significantly advanced our understanding and management of genetic disorders and cancers.

Advances in FISH technology

Since its introduction, FISH technology has experienced considerable progress, enhancing its functionality and increasing both its efficiency and precision. These developments have transformed FISH into a more robust and adaptable instrument within the field of molecular cytogenetics.

MULTIPLEX FISH (M-FISH) AND SPECTRAL KARYOTYPING (SKY)

Multiplex fluorescence in situ hybridization (M-FISH) is an advanced technique that employs a variety of probes, each tagged with distinct fluorophores, to concurrently visualize multiple chromosomal regions. By integrating different fluorophores, M-FISH enables the differentiation of all 24 human chromosomes within a single experimental setup. This capability facilitates an in-depth examination of intricate chromosomal rearrangements and the detection of subtle translocations that conventional karyotyping might overlook (Shen et al., 2022).

Spectral karyotyping (SKY) shares similarities with M-FISH, as it also utilizes a range of fluorescent probes to assign unique colors to each chromosome. This method requires specialized imaging technology that captures the complete spectrum of fluorescence emitted by the probes. SKY offers a comprehensive karyotype of the entire genome, thereby enhancing the identification of complex chromosomal abnormalities and cryptic translocations that traditional techniques may fail to reveal (Shen et al., 2022).

AUTOMATION AND DIGITAL IMAGING

The advent of automated FISH systems has markedly enhanced both the efficiency and reliability of FISH analyses. These systems are capable of simultaneously processing numerous samples, executing hybridization and washing procedures with a high degree of accuracy, and producing images of exceptional resolution. By minimizing the variability associated with manual handling, automation significantly bolsters the reproducibility of the findings (Lebedev, 2023). Progress in digital imaging technology has substantially improved the visualization and assessment of FISH signals. The use of high-resolution cameras in conjunction with advanced imaging software facilitates the accurate capture and quantification of fluorescent signals. Furthermore, image analysis software can autonomously detect and enumerate FISH signals, thereby streamlining the analysis process and enhancing the precision of the results (Lebedev, 2023).

INTEGRATION WITH OTHER TECHNIQUES

The integration of FISH with other molecular methodologies, including CGH and NGS, has significantly enhanced its diagnostic capabilities. By merging FISH with CGH, researchers can simultaneously identify structural variations and alterations in copy number, thereby facilitating a more thorough examination of chromosomal irregularities. Furthermore, the combination of FISH with NGS allows for the detection of sequence-level mutations alongside structural changes, thus providing a comprehensive genetic characterization of the sample (Chai et al., 2023).

Recent innovations have led to the development of CRISPR-based FISH (CRISPR-FISH), which employs CRISPR-Cas9 technology to precisely target specific genomic sequences. This novel approach improves the specificity and resolution of FISH by allowing for the detection of smaller genomic regions that traditional probes may overlook. The potential applications of CRISPR-FISH are promising, particularly in the realms of research and clinical diagnostics, where it can be instrumental in identifying subtle genetic modifications (Chai et al., 2023).

Clinical and research applications of FISH: case studies

FISH is a pivotal technique employed in both clinical and research environments, renowned for its accuracy and dependability in identifying genetic anomalies.

In the realm of NIPT, the integration of FISH with NGS has revolutionized the detection of chromosomal abnormalities by analyzing fetal DNA present in maternal circulation. This method allows for early identification of potential genetic issues during pregnancy without the associated risks of invasive techniques such as amniocentesis. For instance, in a scenario where standard screening suggested a heightened risk for Down syndrome, FISH analysis of fetal cells validated the presence of trisomy 21, thereby enabling timely genetic counseling and informed decision-making (Xu et al., 2023).

FISH also plays a critical role in cancer diagnostics by identifying chromosomal irregularities. In a notable case involving a 45-year-old male diagnosed with CML, FISH successfully detected the BCR-ABL1 fusion gene, a consequence of the Philadelphia chromosome translocation. This finding not only confirmed the diagnosis but also guided the targeted therapy with tyrosine kinase inhibitors, which significantly enhanced the patient's prognosis (Shen et al., 2022). Additionally, in breast cancer assessments, FISH was instrumental in identifying HER2 gene amplification, allowing for the implementation of personalized treatment strategies that improved clinical outcomes.

Moreover, FISH is used for the detection of genetic disorders, such as DiGeorge syndrome, which is marked by a deletion on chromosome 22q11.2. In a case involving a 3-year-old boy presenting with congenital heart defects and developmental delays, FISH confirmed the deletion at 22q11.2, thereby facilitating appropriate medical management and genetic counseling (Liu et al., 2023). Finally, FISH proves effective in the detection of mosaicism, a condition where certain cells exhibit distinct genetic compositions. In a case study of a 15-year-old girl diagnosed with Turner syndrome, FISH successfully identified the mosaic pattern, underscoring its utility in genetic analysis.

Comparative genomic hybridization

CGH is an advanced molecular cytogenetic method that facilitates the detection of CNVs throughout the genome. These CNVs, encompassing both duplications and deletions of extensive DNA segments, can profoundly influence gene expression and functionality, thereby playing a significant role in the etiology of various genetic disorders and malignancies. One of the primary advantages of CGH is its ability to conduct a thorough examination of these genetic variations without necessitating the cultivation of cells, which enhances its efficacy as a tool for genome-wide analysis. The technique involves contrasting the genomic compositions of a test sample, often derived from tumor DNA, with that of a reference sample, usually normal DNA. Both types of DNA are labeled with distinct fluorescent dyes, commonly green for the test sample and red for the reference sample, before being co-hybridized to a target DNA, which may consist of metaphase chromosomes or a microarray with immobilized DNA segments. The underlying mechanism of CGH is predicated on the competitive hybridization between the differently labeled DNAs and the target. Detection and quantification of the fluorescent signals emanating from these hybridized strands yield a ratio of test to reference DNA at each specific genomic locus, allowing for the identification of deviations from the anticipated 1:1 ratio, which signifies the presence of duplications or deletions within the test sample.

Process

The process of DNA extraction involves isolating genomic DNA from both the experimental sample, such as tumor tissue, and a control sample composed of normal tissue.

Upon extraction, the DNA is subjected to labeling with fluorescent dyes; specifically, a green dye is assigned to the test DNA, while a red dye designates the reference DNA. Following labeling, the two DNA samples undergo co-hybridization with a specific target, which may consist of metaphase chromosomes or a microarray. During this hybridization phase, the DNA is denatured to achieve a single-stranded state, facilitating the binding of the labeled DNA to its complementary sequences on the target DNA.

Subsequent to the hybridization, a washing step is performed to eliminate any DNA that has not successfully bound or has bound non-specifically. This critical step enhances the specificity of the hybridization by ensuring that only DNA that is genuinely hybridized to the target remains on the slide.

The analysis of the hybridization is carried out through detection of the fluorescence signals using either a fluorescence microscope or a microarray scanner. By measuring the intensity of these signals, the ratio of test to reference DNA can be calculated for each genomic location. This ratio serves as an indicator of the relative copy number of DNA sequences present in the test sample.

Role in identifying copy number variations (CNVs)

CGH demonstrates a notable efficacy in the detection of CNVs, which are crucial for elucidating the genetic basis of diseases and malignancies. These CNVs can range over extensive portions of DNA, potentially including several genes, thereby influencing gene dosage and resulting in a spectrum of phenotypic outcomes.

In genetic disorders

CGH serves as a crucial tool for identifying microdeletions and microduplications that conventional karyotyping often overlooks. A notable example is DiGeorge syndrome, which arises from a deletion at chromosome 22q11.2. Traditional karyotyping may fail to detect such subtle deletions; however, CGH excels in pinpointing these variations with remarkable accuracy, thereby facilitating precise diagnoses and effective clinical interventions (Liu et al., 2023). Furthermore, CGH is adept at uncovering duplications in genes linked to conditions such as autism spectrum disorders and intellectual disabilities, thereby enhancing our understanding of the genetic underpinnings of these intricate disorders. CNVs are significantly implicated in conditions like autism and intellectual disabilities. Research indicates that specific deletions and duplications within certain genomic regions can influence the manifestation of these disorders. For instance, a duplication of the 15q11-q13 region is correlated with autism, whereas a deletion in the same area may lead to Prader-Willi syndrome or Angelman syndrome. The ability to detect these CNVs through CGH not only aids in achieving more accurate diagnoses but also enriches our comprehension of the genetic mechanisms that contribute to these complex conditions.

In cancer

Comprehensive genomic hybridization has become a vital tool in the field of cancer genomics for the detection of CNVs that are linked to the initiation and advancement of tumors. Cancer cells frequently present with intricate genetic modifications, including both oncogene amplifications and tumor suppressor gene deletions. A notable example is the amplification of the MYC oncogene, which is prevalent across various cancer types and correlates with aggressive tumor characteristics and unfavorable clinical outcomes, as highlighted by Shen et al. (2022). Through CGH, both researchers and clinicians are able to pinpoint these specific genetic alterations, thereby enhancing prognostic assessments and aiding in the formulation of therapeutic strategies. Moreover, CGH is instrumental in the identification of potential therapeutic targets in oncological research. By delineating the CNVs present within tumor genomes, researchers can uncover genes that are recurrently affected by amplification or deletion across particular cancer categories. Such genes can prove to be promising candidates for innovative treatment modalities. A pertinent example includes the HER2 gene, whose amplification in breast cancer has facilitated the creation of HER2-targeted therapies, ultimately resulting in substantial improvements in the prognosis for patients diagnosed with HER2-positive breast cancer.

Additionally, CGH contributes significantly to elucidating the clonal evolution of tumors. Tumors frequently consist of heterogeneous cellular populations, each exhibiting unique genetic characteristics. By examining the CNVs present in various tumor regions, CGH offers valuable insights into the genetic diversity inherent within the tumor and the evolutionary mechanisms that underlie tumor development and progression.

Technological improvements in CGH

Since its introduction, CGH technology has experienced considerable progress, leading to enhancements in its resolution, sensitivity, and range of applications (**Table 9.2**). These advancements have broadened the functional scope of CGH, establishing it as a robust instrument for the genome-wide examination of CNVs in both research and clinical environments.

Array-based CGH

Integration of array technology with CGH

Array-based CGH marks a notable improvement over conventional metaphase CGH by merging microarray technology with CGH principles, thereby facilitating enhanced resolution and a more thorough genomic analysis (**Figure 9.1**). This integration has revolutionized genome-wide assessments, allowing for the precise identification of CNVs throughout the genome. In aCGH, a microarray platform is utilized, where DNA probes are affixed to a solid substrate, such as a glass slide. Each individual spot on the array corresponds to specific genomic regions, and the test and reference DNA samples are labeled with distinct fluorescent dyes—commonly green for the test sample and red for the reference—before

Table 9.2 Technological advancements in molecular cytogenetics

Advancement	*Features*	*Impact on molecular diagnostics*
Multiplex FISH (M-FISH)	Uses multiple fluorophores for simultaneous detection of multiple chromosomal regions	Comprehensive analysis of complex chromosomal rearrangements, improved detection of subtle translocations (Shen et al. 2022)
Array-based CGH (aCGH)	Integrates microarray technology with CGH for higher resolution	Enhanced detection of small CNVs, comprehensive genome-wide analysis, targeted arrays for specific conditions (Lebedev 2023)
NGS-integrated CGH	Combines CGH with NGS for detailed characterization of CNVs and sequence-level mutations	High sensitivity and specificity, dual detection of structural and sequence variations, improved diagnostic accuracy (Chai et al. 2023)
CRISPR-based FISH	Uses CRISPR-Cas9 for high-precision targeting of genomic sequences	Increased specificity and resolution, potential for detecting subtle genetic alterations (Chai et al. 2023)

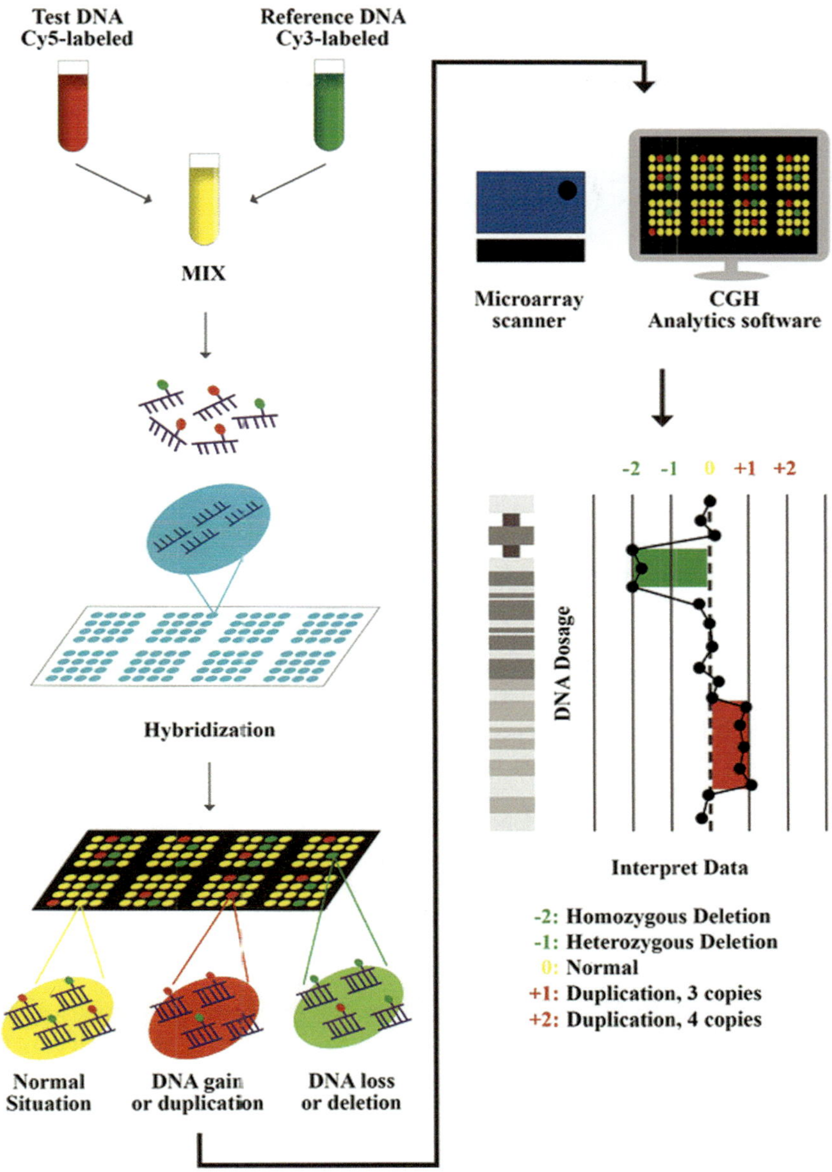

Figure 9.1 Array-based CGH illustrates labelled test and reference DNA co-hybridized on a microarray to detect copy number variations through color-coded signal analysis.

being co-hybridized onto the array. The resulting fluorescent signals are then detected and quantified, yielding a detailed high-resolution map of CNVs across the genome.

The shift from metaphase CGH to array-based CGH has markedly enhanced the resolution of CNV detection. Traditional metaphase CGH was limited to a resolution of approximately 5–10 megabases (Mb), whereas aCGH arrays can accommodate densely arranged probes that enable the detection of CNVs down to kilobase (kb) levels (Shen et al., 2022). This heightened resolution allows for the identification of smaller CNVs that may remain undetected by older methodologies. Furthermore, array-based CGH offers extensive genome-wide coverage, ensuring the accurate detection of CNVs across the entire genome. This feature is particularly advantageous for uncovering novel genetic variations linked to complex traits and diseases. The capability to design custom arrays that focus on specific genomic regions of interest further amplifies the applicability of aCGH in both clinical and research environments. The procedure initiates with the extraction of genomic DNA from both the test and reference samples, followed by the labeling of the DNA with unique fluorescent dyes. The labeled DNA samples are then co-hybridized.

Clinical diagnostic applications

The improved resolution and diagnostic precision of aCGH have resulted in its extensive utilization across numerous clinical diagnostic settings, offering essential understanding of genetic disorders and malignancies.

PRENATAL DIAGNOSTICS

Array-based CGH has become a pivotal tool in prenatal diagnostics, enabling the early detection of chromosomal abnormalities during pregnancy. This technique is particularly adept at identifying CNVs linked to prevalent aneuploidies, such as trisomy 21, which is associated with Down syndrome, as well as smaller chromosomal alterations that can lead to conditions like DiGeorge syndrome (22q11.2 deletion) and Williams syndrome (7q11.23 deletion). The capacity to recognize these genetic anomalies at an early stage is crucial, as it allows for timely medical intervention and management, thereby enhancing clinical outcomes (Xu et al., 2023). In the realm of prenatal diagnostics, aCGH is frequently employed to examine cell-free fetal DNA extracted from maternal blood. This non-invasive technique, referred to as NIPT, offers a safe and precise means of detecting chromosomal abnormalities, circumventing the risks associated with invasive methods such as amniocentesis. By uncovering CNVs that are related to congenital anomalies, aCGH not only supports early diagnosis but also enables appropriate clinical management and genetic counseling, ultimately improving patient care (Xu et al., 2023).

CANCER GENOMICS

In the field of oncology, aCGH is instrumental in revealing genetic changes that underlie tumor initiation and advancement. For example, aCGH is capable of identifying amplifications of oncogenes like HER2 within breast cancer cases, thereby influencing the administration of targeted treatments, such as trastuzumab. Furthermore, it can detect deletions of critical tumor suppressor genes, including TP53, which contributes to a deeper understanding of tumor dynamics and patient prognosis. The extensive genome-wide assessments facilitated by aCGH allow for the discovery of novel genetic markers and potential therapeutic targets, paving the way for tailored cancer therapies (Shen et al., 2022). Additionally, aCGH is utilized to examine the genetic heterogeneity present within tumors. Given that tumors frequently consist of genetically varied cellular populations, deciphering this heterogeneity is vital for anticipating treatment responses and the likelihood of resistance. By pinpointing CNVs across various sections of a tumor, aCGH sheds light on the clonal evolution of the malignancy and aids in crafting more efficacious treatment approaches (Lebedev, 2023).

NEURODEVELOPMENTAL DISORDERS

Array-based CGH plays a crucial role in the diagnosis of neurodevelopmental disorders associated with CNVs. Notably, specific duplications and deletions in chromosomal regions such as 15q11-q13 and 16p11.2 have been correlated with conditions like autism spectrum disorders and intellectual disabilities. The identification of these CNVs through aCGH enhances our understanding of the genetic underpinnings of these intricate disorders, thereby facilitating the development of targeted therapeutic strategies and effective genetic counseling approaches (Lebedev, 2023).

In the context of intellectual disabilities and developmental delays, aCGH is instrumental in detecting CNVs that may influence the clinical presentation. For instance, a duplication of the MECP2 gene located on the X chromosome can lead to MECP2 duplication syndrome, which is characterized by intellectual disability, motor impairments, and seizures. The ability to identify such CNVs not only enables accurate diagnosis but also informs appropriate clinical management, ultimately enhancing the quality of life for individuals affected by these conditions and their families (Shen et al., 2022).

CARDIOVASCULAR DISEASES

Recent studies have highlighted the significance of CNVs in the context of cardiovascular diseases. Utilizing aCGH allows for the detection of genetic variations that are associated with congenital heart defects and various forms of cardiomyopathy. Notably, CNVs affecting the GATA4 gene have been correlated with congenital heart disease. The identification of these genetic changes is crucial for facilitating early diagnosis, assessing risk, and developing personalized treatment approaches, ultimately enhancing patient outcomes (Xu et al., 2023). In instances of unexplained cardiomyopathies, array CGH proves valuable in uncovering CNVs that may contribute to the disease. For example, alterations such as deletions or duplications in the MYH7 gene, which encodes the β-myosin heavy chain, can lead to conditions like hypertrophic or dilated cardiomyopathy. By revealing these CNVs, array CGH offers essential diagnostic insights that can inform clinical management and provide guidance for genetic counseling for patients and their families (Lebedev, 2023).

RARE GENETIC DISORDERS

Array-based CGH has proven to be particularly valuable in the identification of rare genetic disorders linked to CNVs. This technique is adept at uncovering microdeletions and microduplications that contribute to syndromic conditions, including Prader-Willi syndrome and Angelman syndrome. The enhanced resolution and precision offered by aCGH facilitate accurate diagnostic insights, which are crucial for effective medical management and genetic counseling for those impacted and their families (Shen et al., 2022). In instances where traditional karyotyping does not reveal a genetic basis for a clinical presentation, aCGH can deliver a more comprehensive evaluation. For instance, it is capable of identifying the deletion in the 17p13.3 region that is linked to Miller-Dieker syndrome, a disorder marked by lissencephaly and significant developmental delays. The detection of such CNVs is instrumental in establishing a conclusive diagnosis and ensuring appropriate clinical interventions, thereby enhancing the prognosis for affected individuals (Lebedev, 2023).

Next-generation sequencing (NGS)-integrated CGH

NGS-integrated CGH marks a pivotal evolution in the domain of molecular cytogenetics. This cutting-edge methodology synergizes the extensive sequencing capabilities of NGS with the thorough genome-wide assessment provided by CGH. The amalgamation of these technologies facilitates a meticulous and precise analysis of genetic variations, yielding valuable insights into both structural and sequence-level modifications throughout the genome.

The NGS-integrated CGH process encompasses several stages that capitalize on the advantages of both methodologies. Initially, array-based CGH is employed to pinpoint genomic regions exhibiting CNVs, offering a broad perspective of the genomic architecture and identifying areas that require deeper exploration. Subsequently, the regions identified are subjected to sequencing via NGS, which delivers base-pair resolution of the CNVs and uncovers additional genetic alterations, including single-nucleotide variants (SNVs) and small insertions/deletions (indels). The procedure commences with the extraction of genomic DNA from both the test and reference samples, which are then labeled with distinct fluorescent dyes—green for the test DNA and red for the reference DNA. These labeled DNAs are co-hybridized onto a microarray platform, where DNA probes are affixed to a solid substrate. During the hybridization phase, the labeled DNAs competitively attach to their complementary sequences on the probes. The resulting fluorescent signals are captured and quantified using a fluorescence scanner, generating a comprehensive map of CNVs across the genome. This map serves as a critical foundation for the subsequent NGS analysis, which refines the CNVs and delineates their precise breakpoints. The integration of NGS with CGH markedly improves the sensitivity and specificity of CNV detection, as NGS offers base-pair resolution that allows for the identification of smaller CNVs that might be overlooked by array-based CGH alone. This heightened resolution ensures the detection of even low-level mosaicism and subtle genomic alterations.

Applications in clinical diagnostics

The increased sensitivity and specificity of NGS integrated with CGH have resulted in its extensive utilization across numerous clinical diagnostic settings, offering comprehensive insights into genetic diseases and malignancies.

CANCER GENOMICS

NGS-integrated CGH plays a crucial role in the field of cancer genomics by enabling the identification of intricate genetic modifications that contribute to tumor initiation and advancement. For example, in the context of breast cancer, this methodology is adept at revealing amplifications of oncogenes such as HER2, alongside mutations in critical tumor suppressor genes like TP53. The comprehensive genomic analysis facilitated by NGS-integrated CGH supports the formulation of tailored treatment strategies that are informed by the unique genetic landscape of an individual patient's tumor (Shen et al., 2022). Moreover, beyond the detection of established oncogenic alterations, NGS-integrated CGH is instrumental in uncovering novel genetic variations linked to cancer. By offering an extensive overview of the tumor genome, this technique aids in pinpointing potential therapeutic targets and biomarkers that could inform innovative treatment approaches. The capacity to identify both CNVs and sequence-level mutations is essential for grasping the genetic diversity present within tumors, thereby enhancing the development of effective therapeutic interventions (Lebedev, 2023).

PRENATAL AND POSTNATAL DIAGNOSTICS

NGS-integrated CGH is pivotal in cancer genomics, facilitating the detection of complex genetic alterations that drive tumor development and progression. In breast cancer, for instance, this approach effectively identifies amplifications of oncogenes such as HER2 and mutations in essential tumor suppressor genes like TP53. The in-depth genomic insights provided by NGS-integrated CGH are crucial for devising personalized treatment plans that reflect the specific genetic characteristics of a patient's tumor (Shen et al., 2022). Furthermore, in addition to recognizing known oncogenic changes, NGS-integrated CGH plays a significant role in discovering novel genetic alterations associated with cancer. By delivering a comprehensive analysis of the tumor genome, this technique helps identify potential therapeutic targets and biomarkers that could lead to the development of new treatment modalities. The ability to detect both CNVs and sequence-level mutations is vital for understanding the genetic heterogeneity within tumors, thereby contributing to the advancement of effective therapeutic strategies (Lebedev, 2023).

NEURODEVELOPMENTAL AND PSYCHIATRIC DISORDERS

The utilization of NGS combined with CGH in the study of neurodevelopmental and psychiatric disorders has yielded significant understanding of the genetic underpinnings of these ailments. This methodology is particularly effective in detecting CNVs linked to conditions such as autism spectrum disorders, intellectual disabilities, and schizophrenia. The comprehensive genetic data acquired through this technique contributes to a clearer understanding of the intricate genetic framework of these disorders, thereby informing the creation of targeted therapeutic strategies and interventions (Shen et al., 2022).

INHERITED GENETIC DISORDERS

NGS integrated with CGH proves to be a valuable tool in the diagnosis of inherited genetic disorders linked to CNVs and sequence-level mutations. This methodology is particularly adept at detecting gene deletions and duplications that are implicated in various conditions, including cystic fibrosis, muscular dystrophy, and hereditary cancer syndromes. The high resolution and extensive scope of this technique guarantee the identification of all pertinent genetic variations, thereby facilitating a thorough genetic diagnosis for individuals affected by these disorders (Chai et al., 2023).

Advanced techniques and innovations

Single-cell sequencing has emerged as a revolutionary advancement in molecular cytogenetics, allowing researchers to analyze genetic data with an unprecedented level of detail. This cutting-edge technology enables the investigation of the genome, transcriptome, and epigenome at the level of individual cells, uncovering insights into cellular diversity, developmental processes, and the mechanisms of diseases that are often hidden in bulk sequencing methods. The integration of single-cell sequencing into molecular diagnostics has significantly enhanced our understanding of genetic disorders, various cancers, and complex biological systems. The single-cell sequencing process involves isolating individual cells, extracting their genetic material, and performing high-throughput sequencing to create detailed genomic, transcriptomic, or epigenomic profiles. A variety of single-cell sequencing techniques and platforms have been developed, each tailored to address specific genetic research questions. One prominent method is single-cell RNA sequencing (scRNA-seq), which analyzes the transcriptome of individual cells to capture distinct gene expression profiles. Commonly used techniques include Smart-seq2, Drop-seq, and 10x Genomics Chromium. Smart-seq2 is known for its high sensitivity and comprehensive transcript coverage, making it particularly effective for identifying alternative splicing and isoform variations. In contrast, Drop-seq and 10x Genomics Chromium enable high-throughput analyses by processing thousands of cells simultaneously, using microfluidics to encapsulate single cells in droplets for barcoded RNA capture. Additionally, single-cell DNA sequencing (scDNA-seq) focuses on sequencing the entire genome of individual cells, utilizing methods such as multiple displacement amplification and multiple annealing and looping-based amplification cycles (MALBAC) to amplify DNA from single cells, ensuring the high coverage and precision necessary for detecting genetic variations.

Applications in understanding genetic heterogeneity and mosaicism

Single-cell sequencing has profoundly enhanced our comprehension of genetic variability and mosaicism present in tissues, uncovering the array of cell types, their functional conditions, and their evolutionary paths in both healthy and pathological scenarios.

Genetic heterogeneity

Genetic heterogeneity denotes the presence of varied genetic profiles within a cellular population, serving as a fundamental characteristic of numerous biological phenomena, such as development, differentiation, and the progression of diseases. The advent of single-cell sequencing technology has facilitated an unparalleled examination of this heterogeneity, allowing for detailed characterization at an individual cell level.

In the realm of cancer research, single-cell sequencing has illuminated the genetic variability present within tumors, uncovering distinct subpopulations of cancer cells that possess unique genetic mutations and gene expression patterns. This genetic diversity is closely linked to critical processes such as tumor progression, metastasis, and resistance to treatment. For instance, studies utilizing single-cell RNA sequencing in breast cancer have identified various subpopulations of cells, each exhibiting different metastatic capabilities and responses to therapeutic interventions. Furthermore, tumors frequently display clonal diversity, with multiple subclones containing specific genetic alterations that influence the tumor's behavior and treatment efficacy. By analyzing individual cells, researchers can pinpoint rare subclones that contribute to therapeutic resistance or facilitate metastasis, thereby informing the development of personalized treatment approaches.

Additionally, single-cell sequencing has proven instrumental in exploring the diversity of immune cell populations. This technique has revealed significant fluctuations in immune cell composition and gene expression during immune responses and across various diseases, providing essential insights into immune system functionality and the advancement of immunotherapies. By tracking the differentiation and activation states of immune cells, researchers can discern how different subsets respond to pathogens, leading to the identification of distinct populations of T cells, B cells, and myeloid cells, each playing specific roles in immune responses. Such knowledge is pivotal for the formulation of targeted immunotherapies and the design of effective vaccines.

Mosaicism

Mosaicism is characterized by the existence of genetically distinct populations of cells within a single organism, arising from mutations that take place post-fertilization. The technique of single-cell sequencing is particularly effective in identifying and analyzing mosaicism, as it can detect rare mutations that may be present in only a small proportion of cells.

During the developmental phase, somatic mutations can lead to the emergence of mosaicism. Research utilizing single-cell sequencing on human brain tissues has uncovered somatic mutations within neuronal cells, offering valuable insights into the genetic variability and evolutionary processes of the brain (Lodato et al., 2015). The brain displays significant levels of mosaicism, with individual neurons containing distinct somatic mutations. This genetic heterogeneity may play a role in the variations observed in brain function and the predisposition to neurological disorders. By examining single neurons, scientists can elucidate the lineage and clonal relationships among cells in the brain, thereby enhancing our understanding of the developmental mechanisms underlying mosaicism.

Mosaicism is also associated with a range of genetic disorders and cancers. For instance, single-cell DNA sequencing of blood cells has revealed mosaic mutations linked to hematological malignancies and clonal hematopoiesis of indeterminate potential (CHIP), a condition that elevates the risk of blood cancer development (Jaiswal et al., 2014). In the context of cancer, mosaic mutations can facilitate clonal expansion and tumor formation. The application of single-cell sequencing enables the early detection of these mutations, which is crucial for the timely diagnosis and monitoring of disease progression. This methodology is similarly employed to investigate mosaicism in genetic disorders, where mutations may only be present in a fraction of cells, resulting in diverse clinical manifestations.

Case studies in cancer diagnostics and prenatal screening

Single-cell sequencing has been applied in numerous case studies, highlighting its potential in cancer diagnostics and prenatal screening.

Cancer diagnostics

Breast cancer

A pivotal investigation employed single-cell RNA sequencing to explore the cellular makeup and gene expression patterns within breast cancer tumors. This research uncovered various subpopulations of cancer cells, each exhibiting distinct metastatic capabilities. Through the analysis of individual cells, the study identified a specific group of cells that expressed genes linked to epithelial-mesenchymal transition (EMT), a mechanism associated with the spread of cancer. The implications of these findings are profound, particularly in the context of creating targeted therapies that can effectively address the heterogeneity observed within tumors (Patel et al., 2014). The investigation demonstrated that breast cancer tumors consist of diverse cellular populations, each characterized by unique gene expression profiles. By pinpointing subpopulations associated with EMT, researchers are positioned to formulate therapies aimed at these particular cells, which may help in thwarting metastasis and enhancing patient prognoses. Furthermore, single-cell sequencing has the potential to reveal rare resistant clones that may arise during treatment, thereby informing the design of combination therapies intended to counteract resistance.

Glioblastoma

In the context of glioblastoma, a notably aggressive form of brain cancer, single-cell RNA sequencing has uncovered significant intratumoral heterogeneity, particularly highlighting the existence of cancer stem cells. These stem cells are characterized by their resistance to standard treatment modalities, which plays a crucial role in the recurrence of the tumor. By analyzing these cells at the single-cell level, researchers are able to pinpoint potential therapeutic targets aimed at eradicating cancer stem cells, thereby enhancing treatment efficacy and patient outcomes. The findings indicate that glioblastoma tumors are organized in a cellular hierarchy, with cancer stem cells positioned at the top. These cells possess unique gene expression profiles and demonstrate heightened resistance to both chemotherapy and radiation therapies. Consequently, the development of targeted therapies aimed at these specific cells could significantly reduce the likelihood of tumor recurrence and improve overall survival rates for patients. The application of single-cell sequencing technology offers the necessary resolution to effectively identify and characterize these vital subpopulations within the tumor microenvironment.

Leukemia

Single-cell DNA sequencing of acute myeloid leukemia (AML) specimens has elucidated the clonal evolution of neoplastic cells. This methodology has facilitated the identification of subclonal populations characterized by unique mutations and varying resistance to pharmacological interventions. A comprehensive understanding of the clonal structure of AML is instrumental in forecasting disease advancement and customizing treatment approaches tailored to individual patients (Navin et al., 2011).

The investigation demonstrated that AML tumors consist of a heterogeneous assembly of subclones, each exhibiting distinct genetic modifications. By analyzing individual cells, researchers are able to pinpoint infrequent subclones that contribute to therapeutic resistance or accelerate disease progression. This insight paves the way for the formulation of targeted therapies that specifically address the genetic variations present in each subclone, thereby enhancing treatment efficacy and minimizing the risk of relapse.

Prenatal screening

Non-invasive prenatal testing (NIPT)

Single-cell sequencing has been utilized in NIPT to examine fetal cells extracted from maternal blood. This innovative technique facilitates the identification of chromosomal abnormalities, including trisomy 21 (Down syndrome), without resorting to invasive procedures such as amniocentesis. By analyzing individual fetal cells, NIPT enables precise and early identification of genetic disorders, thereby enhancing prenatal care and informing decision-making processes (Xu et al., 2023). The application of single-cell sequencing in NIPT offers a detailed perspective of the fetal genome, allowing for the detection of smaller CNVs and mosaicism that may elude conventional screening methods. This methodology significantly improves the accuracy and dependability of prenatal assessments, presenting a safer and less invasive alternative to traditional diagnostic approaches.

Optical genome mapping

OGM represents a sophisticated cytogenetic methodology that offers an extensive and precise visualization of the genome. This technique excels in identifying structural variations and large genomic rearrangements that conventional cytogenetic approaches often struggle to detect (**Figure 9.2**). The following discussion will delve into the foundational principles of OGM, elucidating its advantages over traditional cytogenetic techniques while emphasizing its applications and significance in the realm of molecular diagnostics.

The process of OGM initiates with the extraction of high molecular weight (HMW) DNA from biological samples, a crucial step that ensures the integrity of long DNA molecules necessary for accurate genomic mapping. Unlike conventional DNA extraction methods that tend to fragment DNA, OGM employs specialized protocols designed to maintain the length of DNA strands, thereby facilitating a thorough analysis. Following the extraction, specific DNA sequences are tagged with fluorescent dyes that bind to designated nucleotide sequences, resulting in a unique fluorescent pattern or barcode along the DNA. This labeling is essential, as it generates the distinctive fluorescent signatures that will be visualized and scrutinized throughout the mapping process.

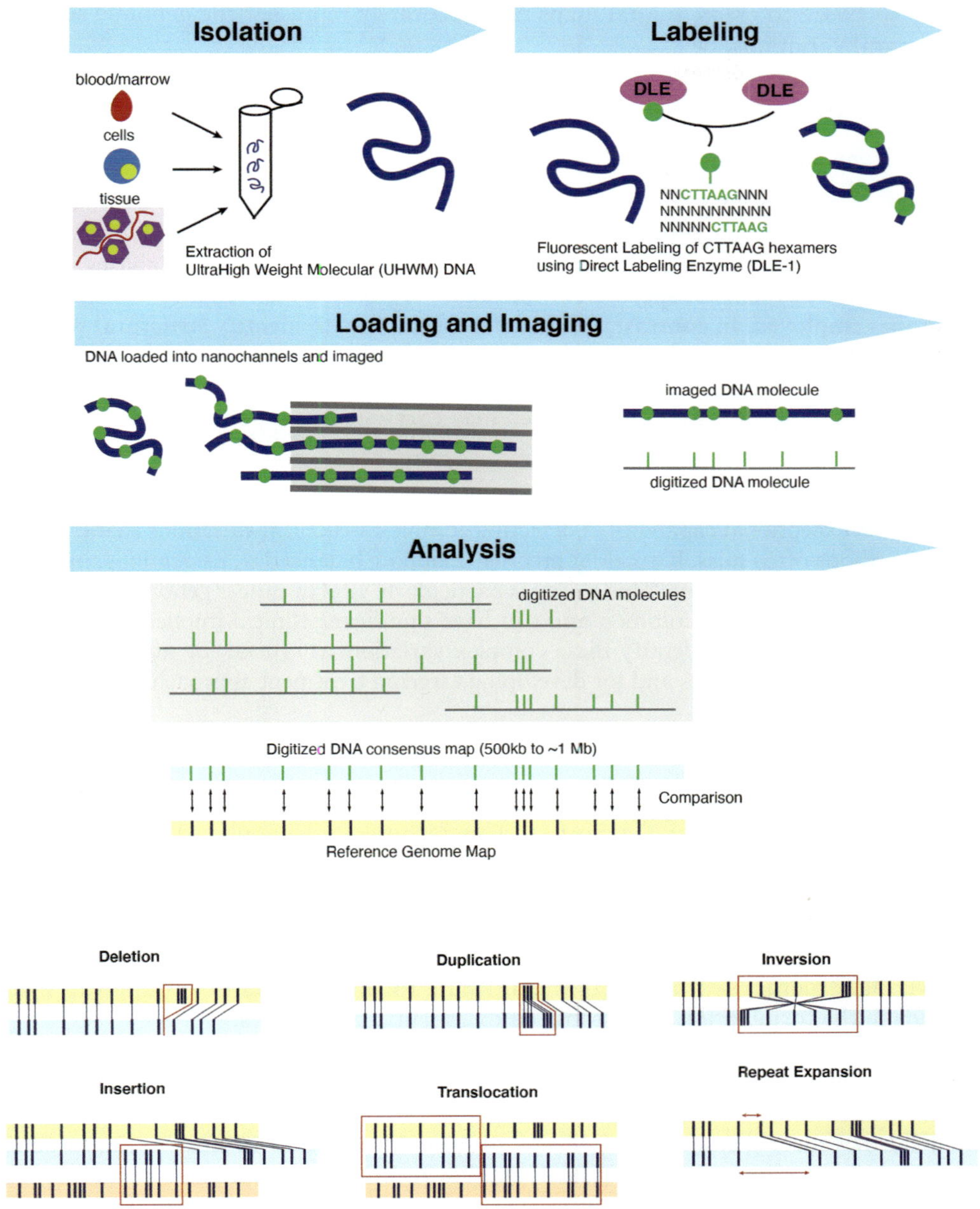

Figure 9.2 Optical genome mapping outlines the process of DNA isolation, fluorescent labeling, nanochannel imaging, and analysis to identify structural variations.

Subsequently, the labeled DNA molecules undergo linearization and stretching on a specialized surface or within nanochannels, a vital procedure that organizes the DNA into an extended form conducive to precise imaging and analysis. Nanochannel arrays are particularly effective in this context, as they provide a controlled environment for the alignment and stretching of DNA molecules. High-resolution fluorescence microscopy is then utilized to capture images of the labeled DNA, where the fluorescent markers yield a barcode-like pattern indicative of the specific sequences present. These images are processed through advanced software to create digital maps that accurately represent the genomic sequences and structural variations.

Benefits over traditional cytogenetics

Higher resolution and sensitivity

Traditional cytogenetic methodologies, including karyotyping and FISH, exhibit notable constraints regarding their resolution and sensitivity. Karyotyping is capable of identifying only substantial chromosomal alterations, typically exceeding 5–10 megabases, whereas FISH, while able to focus on specific chromosomal regions, is restricted by the dimensions of the probes employed. In contrast, OGM offers the ability to identify structural variations as minor as a few kilobases, thereby achieving considerably enhanced resolution and sensitivity (Chan et al., 2018). This increased resolution facilitates the identification of smaller and more nuanced genetic alterations, which may carry significant clinical relevance.

Improved detection of complex variations

OGM exhibits exceptional capability in detecting complex structural variations that traditional detection methods often miss. It excels at precisely mapping balanced translocations, inversions, and large insertions or deletions. This ability is particularly vital in cancer genomics, where such intricate rearrangements are common and can have significant clinical implications (Goldrich et al., 2021). The ability to identify these complex variations is crucial for understanding the genetic foundations of diseases and for developing targeted treatment approaches.

Applications in molecular diagnostics

The exceptional resolution and extensive capabilities of OGM render it an essential instrument in the field of molecular diagnostics. This technology is employed to detect genetic irregularities associated with a range of conditions, such as cancers, hereditary disorders, and congenital malformations. OGM offers significant insights into the genetic underpinnings of various diseases, thereby informing treatment strategies and enhancing patient care (Chan et al., 2021). For example, in the context of cancer diagnostics, OGM is adept at uncovering structural variations that contribute to tumor initiation and advancement, which aids in the formulation of customized therapeutic approaches.

Cancer genomics

In the field of cancer genomics, OGM has proven instrumental in uncovering structural variations that contribute to the initiation and advancement of tumors. A notable application of OGM is its use in identifying intricate rearrangements within hematological cancers, including the translocations associated with the BCR-ABL1 gene in CML. Traditional diagnostic techniques often struggle to detect these complex alterations, yet they are essential for effective diagnosis and therapeutic strategies (Liu et al., 2023). By generating comprehensive maps of these genetic rearrangements, OGM enhances the precision of cancer diagnosis and facilitates improved management of the disease.

Genetic disorders

One of the significant applications of OGM lies in the diagnosis of genetic disorders. It has proven effective in detecting structural variations in individuals experiencing developmental delays and intellectual disabilities. A notable study demonstrated that OGM uncovered a previously unrecognized duplication in a patient exhibiting a complex phenotype, which facilitated a more precise diagnosis and customized intervention (Mantere et al., 2021). This functionality is especially crucial for identifying conditions characterized by intricate genetic foundations that may elude detection through conventional diagnostic techniques.

Prenatal screening

In the field of prenatal diagnostics, OGM offers a non-invasive approach to detecting chromosomal abnormalities. This method analyzes fetal DNA extracted from maternal blood, allowing for the early and accurate identification of conditions such as trisomy 21 (Down syndrome), all while avoiding the risks associated with invasive techniques like amniocentesis (Zhang et al., 2020). The ability to provide detailed genetic information without invasive procedures makes OGM an attractive option for prenatal screening and diagnostic applications.

CONCLUSION

The discipline of molecular cytogenetics has made remarkable strides in the realm of molecular diagnostics through the adoption of advanced methodologies such as FISH, CGH, and NGS. These techniques facilitate high-resolution and extensive genomic evaluations that are essential for the diagnosis, prognosis, and management of various genetic disorders, cancers, and prenatal conditions. The introduction of these technological advancements has significantly improved the detection of subtle genetic variations, thereby enabling more accurate and individualized medical insights. Moreover, innovations like single-cell sequencing and OGM further enhance our ability to decipher intricate genetic landscapes, paving the way for targeted therapeutic strategies and better patient outcomes. As this field progresses, the amalgamation of these cutting-edge techniques is poised to transform our methodologies in genetic research and clinical practice, ultimately improving patient care and propelling the precision medicine movement forward.

REFERENCES

Chai, Y., Zhang, P., & Wang, Y. (2023). Optical genome mapping for detecting complex and cryptic chromosomal rearrangements. Molecular Cytogenetics, 16(29), 5, 12, 20.

Chan, S., Lam, E., Saghbini, M., Bocklandt, S., Hastie, A., Cao, H., Holmlin, E., Borodkin, M. (2018). Structural variation detection and analysis using bionano optical mapping. Methods. Molecular Biology, *1833*, 193–203. https://doi.org/10.1007/978-1-4939-8666-8_16. PMID: 30039375

Goldrich, D.Y., LaBarge, B., Chartrand, S., Zhang, L., Sadowski, H.B., Zhang, Y., Pham, K., Way, H., Lai, C.J., Pang, A.W.C., Clifford, B., Hastie, A.R., Oldakowski, M., Goldenberg, D., Broach, J.R. (2021). Identification of somatic structural variants in solid tumors by optical genome mapping. Journal of Personalized Medicine, *11*(2), 142.

Jaiswal, S., Fontanillas, P., Flannick, J., Manning, A., Grauman, P. V., Mar, B. G., ... Ebert, B. L. (2014). Age-related clonal hematopoiesis associated with adverse outcomes. New England Journal of Medicine, 371(26), 2488–2498.

Lebedev, I. (2023). Advances in molecular cytogenetics. Biomedicines, Special Issue.

Liu, Y., Huang, N., Lu, W., Luo, L., Yuan, H., Pan, L., Ding, S., & Yang, B. (2023). Characteristics and clinical evaluation of X chromosome translocations. Molecular Cytogenetics, 16(36)., 4, 6, 7.

Lodato, M. A., Woodworth, M. B., Lee, S., Evrony, G. D., Mehta, B. K., Karger, A., ... & Walsh, C. A. (2015). Somatic mutation in single human neurons tracks developmental and transcriptional history. Science, 350(6256), 94–98.

Mantere, T., Neveling, K., Pebrel-Richard, C., Benoist, M., van der Zande, G., Kater-Baats, E., ... El Khattabi, L. (2021). Optical genome mapping enables constitutional chromosomal aberration detection. The American Journal of Human Genetics, 108(8), 1409–1422.

Navin, N., Kendall, J., Troge, J., Andrews, P., Rodgers, L., McIndoo, J., ... Hicks, J. (2011). Tumour evolution inferred by single-cell sequencing. Nature, 472(7341), 90–94.

Patel, A. P., Tirosh, I., Trombetta, J. J., Shalek, A. K., Gillespie, S. M., Wakimoto, H., ... Regev, A. (2014). Single-cell RNA-seq highlights intratumoral heterogeneity in primary glioblastoma. Science, 344(6190), 1396–1401.

Xu, Y., Hu, S., Chen, L., Hao, Y., Zhang, H., Xu, Z., Wu, W., & Deng, L. (2023). Application of non-invasive prenatal testing in screening chromosomal aberrations. Molecular Cytogenetics, 16(29), 5, 9, 10, 19.

Chapter 10

Microfluidics and lab-on-a-chip technologies in molecular diagnostics

*Ayyub Ali Patel**

Department of Clinical Biochemistry, College of Medicine,
King Khalid University, Abha, Saudi Arabia

INTRODUCTION

Microfluidics is a technology that involves manipulating small volumes of fluids through microchannels. It plays a significant role in biomedical diagnostics, environmental monitoring, and chemical synthesis (**Figure 10.1**).

Microfluidic devices mostly function in laminar flow conditions as a result of their low Reynolds numbers. This leads to the movement of fluids in a uniform and parallel manner, without any disturbances, enabling accurate manipulation of fluid blending and chemical reactions. The presence of laminar flow is essential in situations where regulated reaction conditions are necessary, such as in the study of enzyme kinetics and cell culture (Whitesides & Stroock, 2001).

At a microscale, surface tension and capillary forces are more important than gravitational and inertial forces, allowing for unique fluid handling capabilities. Capillary action is utilized in paper-based microfluidic devices, which are extensively employed in point-of-care (POC) diagnostics to enable rapid and cost-effective testing. Advancements in material science have resulted in the creation of superhydrophobic and hydrophilic surfaces, which improve the management of fluids in these systems (Yetisen et al., 2013).

The development of microfabrication technologies, including photolithography, soft lithography, and 3D printing, has greatly transformed the manufacturing process of microfluidic devices. These methods facilitate the creation of intricate microchannel networks with exceptional accuracy and consistency. Photolithography and soft lithography are still considered the most reliable methods for creating microfluidic devices with high resolution and multiple functions. 3D printing has recently become a flexible technique for quickly constructing prototypes and complex 3D constructions (Nguyen et al., 2018).

The integration of microfluidics with biosensors has resulted in the development of diagnostic platforms that are both highly sensitive and selective. Electrochemical biosensors exhibit significant potential for the continuous monitoring of biological and chemical analytes in real-time. Current research is centered on enhancing the sensitivity, selectivity, and stability of these biosensors, rendering them well-suited for use in clinical diagnostics, environmental monitoring, and food safety applications (Zheng & Yang, 2023).

Droplet microfluidics entails the creation and control of individual droplets within a continuous medium. This technology enables high-throughput screening and single-cell analysis by encapsulating individual cells or reagents in droplets, allowing parallel processing and analysis. Recent advances in droplet generation techniques have enhanced the control of droplet characteristics like size, composition, and merging. This has resulted in progress in the fields of drug discovery and synthetic biology (Teh et al., 2008).

* Corresponding author.

DOI: 10.1201/9781003489177-10

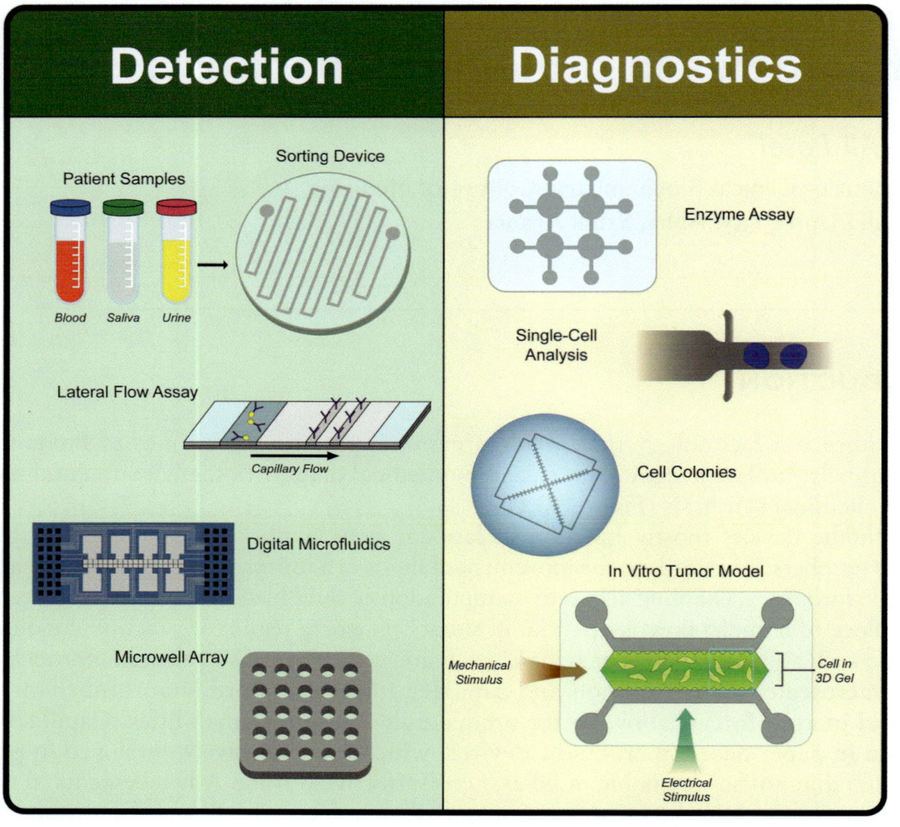

Figure 10.1 Overview of detection and diagnostics technologies using microfluidic and lab-on-a-chip platforms for analyzing patient samples, enzyme activity, cell behavior, and tumor models.

Lab-on-a-chip (LOC) systems combine many laboratory functions, including sample preparation, reaction, separation, and detection, into a single chip. These systems provide notable benefits in terms of mobility, decreased use of reagents, and quicker analysis durations. Current advancements concentrate on improving the integration and automation of LOC systems, with the aim of making them more user-friendly and appropriate for POC applications (Sackmann et al., 2014).

DESIGN AND FABRICATION OF LAB-ON-A-CHIP DEVICES

Fluid dynamics

Fluid dynamics are crucial in establishing the efficiency and functioning of LOC devices. A key factor to take into consideration is the behavior of fluids at the microscale, which is primarily characterized by laminar flow. Laminar flow is a state in which a fluid moves in smooth, parallel layers without any disturbance between them. This behavior is observed at low Reynolds numbers, usually below 2000. This pattern is in contrast to turbulent flow, which is characterized by the prevalence of chaotic eddies and vortices.

Laminar flow guarantees that the blending of fluids takes place mostly through diffusion rather than turbulence. This attribute enables accurate manipulation of the interaction between distinct fluid streams. In certain applications, such as chemical reactions or biological tests, it is important to maintain laminar flow. This ensures that chemicals only mix at the desired locations, which improves the reproducibility and accuracy of the reactions (Stone & Kim, 2001).

In designing microfluidic channels, engineers account for factors such as channel dimensions, flow rates, and fluid properties (viscosity and density). Computational fluid dynamics (CFD) simulations are commonly used to simulate and adjust these parameters, ensuring optimal fluid handling within the LOC device. In addition, the utilization of serpentine or zigzag channel configurations can improve the process of mixing by expanding the surface area of contact between fluids. This helps to overcome the restricted diffusion that occurs in laminar flow conditions (Squires & Quake, 2005).

Materials

The choice of materials for LOC devices is crucial as it affects not only the mechanical and chemical characteristics of the device but also its biocompatibility and optical transparency. Polydimethylsiloxane (PDMS) is a frequently utilized substance because of its exceptional transparency, compatibility with living organisms, and simplicity in creating objects using soft lithography techniques. PDMS is highly preferred for biological applications due to its gas-permeability, which enables cell cultures to undergo respiration within the device (McDonald et al., 2000).

Nevertheless, PDMS possesses several limits, such as its hydrophobic characteristic, which can result in the adsorption of hydrophobic molecules and impact the movement of fluids. PDMS can be rendered hydrophilic through surface treatments, such as oxygen plasma treatment, however this effect is very temporary and decreases over time. Depending on the specific function, alternative materials such as polymethylmethacrylate (PMMA), glass, and thermoplastics are also employed. Glass possesses exceptional chemical resistance and optical characteristics, rendering it well-suited for applications that necessitate high-resolution optical detection (Becker & Gärtner, 2008).

Thermoplastics, whether subjected to injection molding or hot embossing, offer a cost-efficient method for large-scale manufacture of LOC devices. Progress in materials research has also resulted in the creation of new polymers that have specific features, such as increased ability to attract water or improved ability to resist chemicals. This has further broadened the potential uses for LOC devices (Fiorini & Chiu, 2005).

Integration of components

LOC devices are characterized by their capacity to combine various laboratory tasks on a single chip, such as sample preparation, reaction, separation, and detection. This integration necessitates thorough design to guarantee that every component operates in synergy within the compact footprint of the device.

Microfluidic valves and pumps are crucial for regulating fluid flow within the device. Pneumatically actuated valves, fabricated by the process of multilayer soft lithography, enable accurate manipulation of fluid flow without the requirement of external pumps (Unger et al., 2000). Passive capillary valves, which operate using surface tension, can be specifically engineered to open and close depending on the location of the fluid inside the channel.

The detection methods incorporated into LOC devices differ depending on the specific purpose. Optical detection methods, such as fluorescence and absorbance, are widely used because of their high sensitivity and non-intrusive characteristics. These methods frequently necessitate the use of see-through materials and meticulously planned optical routes to optimize the capturing of signals. Conductive materials and accurate electrode positioning are essential for applications that involve electrical detection, such as electrochemical sensors (Dittrich & Manz, 2006).

Scalability and portability

When transitioning LOC technologies from the laboratory to real-world applications, it is important to consider scalability and portability as crucial factors. In order to be suitable for POC diagnostics, devices need to possess durability, user-friendliness, and the ability to function effectively in settings with limited resources. Therefore, it is essential to create portable energy sources, easy-to-use interfaces, and reliable data analysis tools (Chin et al., 2011).

Mass manufacturing methods, such as injection molding for plastic devices or roll-to-roll processing for flexible substrates, are crucial for increasing output on a large scale. These solutions must guarantee uniform quality and performance across substantial quantities of devices. In addition, the incorporation of wireless communication technologies, like Bluetooth or Wi-Fi, can improve the capabilities of portable LOC devices by allowing for the transfer and analysis of data from a distance (Gervais et al., 2011).

FABRICATION METHODS FOR LAB-ON-A-CHIP DEVICES

The fabrication of LOC devices involve several sophisticated techniques, each offering unique advantages and challenges. These methods have evolved significantly over the past decade, enabling the creation of complex, multifunctional microfluidic systems that are integral to modern diagnostics, chemical analysis, and biological research (**Table 10.1**).

Photolithography

Photolithography is a fundamental method in microfabrication that is extensively employed to produce precise patterns on materials like silicon, glass, or polymers. The process

Table 10.1 Comparison of microfluidic fabrication techniques

Fabrication technique	*Principle*	*Advantages*	*Limitations*	*Typical applications*
Photolithography	Light exposure to a photoresist to create patterns	High precision, high resolution	Requires cleanroom, expensive equipment	High-resolution microfluidic devices, silicon chips
Soft lithography	Replica molding using elastomeric materials	Cost-effective, flexible, biocompatible	Hydrophobic, potential adsorption of small molecules	Rapid prototyping, biological applications
3D printing	Layer-by-layer additive manufacturing	Design flexibility, rapid prototyping	Lower resolution compared to photolithography, material variability	Custom microfluidic devices, complex 3D structures
Micro-milling	Precision cutting of rigid materials	High precision, suitable for rigid materials	Time-consuming, tool wear	Metal and ceramic microstructures, prototyping
Laser ablation	Material removal using a high-energy laser	High precision, flexible material compatibility	Thermal damage, limited to certain materials	Fabrication of microchannels, rigid substrates

commences with substrate preparation, wherein the substrate is thoroughly cleansed to guarantee the elimination of any impurities that may hinder future procedures. Next, a light-sensitive substance called photoresist is applied onto the substrate using a spin-coating technique. Spin coating is a process where a small quantity of photoresist is applied to the center of the substrate and then rapidly rotated at high velocities to create a consistent and thin film.

After the application of the photoresist, the substrate is subjected to ultraviolet (UV) light exposure using a photomask. The photomask includes the intended pattern, and the UV exposure modifies the chemical composition of the photoresist in the areas that are exposed. Next, the substrate is immersed in a solution that eliminates the photoresist that has been exposed, therefore exposing the underlying substrate in the specific pattern determined by the photomask. The patterned substrate undergoes an etching procedure, which can be either wet etching using chemical solutions or dry etching using plasma. This process removes the material from the exposed areas, resulting in the formation of microchannels and other features. Ultimately, the residual photoresist is removed, resulting in the presence of etched microstructures.

Photolithography provides great accuracy and detail, making it well-suited for creating complex microfluidic networks. Nevertheless, the process necessitates costly apparatus and the maintenance of a cleanroom environment, which may provide a challenge for smaller laboratories or settings with limited resources (Madou, 2002).

Soft lithography

Soft lithography is a versatile and cost-effective alternative to photolithography, using elastomeric materials like PDMS to create microfluidic structures. The process starts with the creation of a master mold, typically fabricated using photolithography or other microfabrication techniques. This mold contains the negative of the desired microstructures.

PDMS pre-polymer is poured over the master mold and cured at an elevated temperature to solidify. After curing, the PDMS is peeled off from the master mold, replicating the microstructures on its surface. The PDMS replica is then bonded to another PDMS layer or a glass substrate to form enclosed microchannels. Bonding can be achieved through oxygen plasma treatment, which activates the surfaces and enables strong, irreversible bonds.

Soft lithography is particularly advantageous for rapid prototyping and is less expensive than photolithography. It allows for the creation of flexible, biocompatible devices that are well-suited for biological applications. However, PDMS is hydrophobic and can adsorb small hydrophobic molecules, potentially interfering with some assays. Surface treatments can mitigate this issue, although they add complexity to the fabrication process (Xia & Whitesides, 1998).

3D printing

3D printing, also known as additive manufacturing, is a highly effective method for creating LOC devices. It provides exceptional design flexibility and allows for the creation of intricate three-dimensional structures by adding layers one by one. The method commences with creating the microfluidic device using computer-aided design (CAD) software. The 3D printer converts the computer design into a tangible object by adding material in successive layers.

LOC fabrication utilizes several 3D printing technologies, such as stereolithography (SLA), fused deposition modeling (FDM), and inkjet printing. SLA is a process that entails using a UV laser to solidify a photopolymer resin, gradually constructing the desired form by adding one layer at a time. FDM uses a thermoplastic filament that is heated and extruded to create successive layers. Inkjet printing involves the deposition of droplets of material, which are subsequently cured or solidified.

3D printing allows for quick creation of prototypes and customization, making it perfect for developing unique microfluidic devices designed for specific uses. It has the ability to create complex shapes that are challenging to build with conventional techniques. Nevertheless, the level of detail achieved by 3D printers is not currently on par with that of photolithography. Additionally, the materials utilized in 3D printing can differ in terms of their ability to interact with living organisms and their resistance to chemical substances (Nguyen et al., 2018).

MICRO-MILLING AND LASER ABLATION

Micro-milling and laser ablation are methods of subtractive manufacturing that entail the elimination of material from a substrate in order to produce microstructures. These techniques are especially appropriate for inflexible materials like metals, ceramics, and specific types of polymers.

Micro-milling uses precision milling machines equipped with micro-sized cutting tools to carve microchannels and features into the substrate. This process allows for the creation of high-precision features with complex geometries. Laser ablation employs a concentrated high-energy laser beam to eliminate material from the substrate by ablation, resulting in the formation of microchannels and structures. The laser parameters, including power, pulse duration, and focus, are meticulously regulated to attain the intended characteristics.

Both micro-milling and laser ablation are highly precise and flexible techniques that are well-suited for generating features in materials that cannot be used with regular photolithography. Nevertheless, these methods can be laborious and necessitate thorough fine-tuning to avoid problems like tool deterioration in micro-milling or heat-induced damage in laser ablation (Levy et al., 2013).

Bonding techniques

Following the process of microfabrication, bonding techniques are employed to integrate the separate layers and create a fully functional LOC gadget. Efficient bonding guarantees the integrity of the structure and operational effectiveness of the microchannels, preventing any leakage and preserving the accuracy of fluid flow.

Thermal bonding is the process of using heat and pressure to bond together thermoplastic materials. Accurate manipulation of temperature and pressure is necessary to prevent any distortion of the microstructures. Adhesive bonding is a process that involves using adhesives to connect surfaces, which is then followed by curing. This process is simple and direct; however, it can introduce impurities if not well regulated. Plasma bonding, a technique frequently employed for PDMS, entails subjecting the surfaces to plasma treatment in order to activate them. This activation enables a robust and permanent bonding when the surfaces are forced together (Bhattacharyya et al., 2005).

Every bonding process possesses its own benefits and difficulties. Thermal bonding yields robust and long-lasting connections but requires careful control of process parameters. Adhesive bonding is a flexible method; however, it has the potential to introduce chemicals that are undesirable. Plasma bonding is a highly efficient method for bonding PDMS, although it necessitates the use of specialist equipment and can be influenced by environmental factors (Bhattacharyya et al., 2005).

APPLICATIONS OF LAB-ON-A-CHIP TECHNOLOGIES IN MOLECULAR DIAGNOSTICS

LOC technologies have transformed molecular diagnostics by reducing the size and integrating intricate laboratory processes into a single, portable device. These technologies have several benefits, such as decreased sample and reagent volumes, faster processing times, and the possibility of POC testing (**Table 10.2**). This section examines the primary uses of LOC technologies in molecular diagnostics, emphasizing their influence on disease identification, tracking, and individualized healthcare.

Disease detection

LOC devices have greatly enhanced the detection of infectious diseases by facilitating rapid and accurate identification of pathogens. Conventional diagnostic approaches, such as culture-based techniques and polymerase chain reaction (PCR), typically necessitate sophisticated laboratory equipment and time-consuming procedures. On the other hand, LOC technologies have the ability to carry out these tests at a faster pace and with limited equipment. Microfluidic PCR machines, such as those mentioned by Chin et al. (2011), are capable of rapidly amplifying and detecting genetic material from pathogens. This process is significantly faster compared to traditional approaches.

An important use of this technology is in identifying viral diseases, including HIV, influenza, and most recently, COVID-19. LOC devices specifically engineered for nucleic acid amplification tests (NAATs) have the capability to directly identify viral RNA or DNA from samples obtained from patients, yielding results in a matter of minutes to hours. The prompt delivery of results is vital for prompt identification and therapy, especially in situations of widespread disease where a swift reaction is imperative (Liu et al., 2021).

Cancer diagnostics

LOC technologies have significantly contributed to the field of cancer diagnostics. Circulating tumor cells (CTCs) and cell-free DNA (cfDNA) serve as important biomarkers for early

Table 10.2 Applications of lab-on-a-chip (LOC) technologies in molecular diagnostics

Application	*Specific methods*	*Benefits*	*Examples*
Infectious disease detection	Microfluidic PCR, nucleic acid amplification tests (NAATs)	Rapid diagnosis, minimal equipment	COVID-19 testing, HIV detection
Cancer diagnostics	Circulating tumor cell (CTC) isolation, cfDNA analysis	Non-invasive, high sensitivity	Liquid biopsy for cancer detection, monitoring tumor dynamics
Genetic testing	SNP genotyping, sequencing, gene expression analysis	Personalized medicine, early disease detection	Prenatal diagnostics, pharmacogenomics
Point-of-care testing	Integrated microfluidic systems with optical/electrochemical detection	Rapid results, suitable for resource-limited settings	Malaria testing, tuberculosis detection
Protein and cell-based assays	ELISAs, western blots, organ-on-a-chip systems	High sensitivity, physiologically relevant conditions	Biomarker discovery, drug testing, disease modeling

cancer detection, monitoring disease progression, and assessing treatment response. LOC devices can isolate and analyze these biomarkers from blood samples with high sensitivity and specificity.

Microfluidic devices can separate CTCs based on their physical properties, such as size and deformability, or by using immunoaffinity techniques that target specific surface markers. Once isolated, these cells can be subjected to downstream molecular analyses, including genetic sequencing and protein expression profiling, providing valuable insights into the tumor's characteristics and potential therapeutic targets (Nagrath et al., 2007).

Similarly, LOC platforms for cfDNA analysis enable non-invasive liquid biopsies, which are less invasive than traditional tissue biopsies and can be performed more frequently. These devices use microfluidic technologies to enrich and detect cfDNA from blood samples, allowing for the monitoring of genetic mutations and alterations associated with cancer (Wan et al., 2017). LOC equipment have the ability to separate and examine these biomarkers from blood samples with a high level of sensitivity and accuracy.

Microfluidic devices are capable of segregating CTCs by exploiting their physical characteristics, such as size and flexibility, or by employing immunoaffinity methods that focus on particular surface markers. After being separated, these cells can undergo subsequent molecular investigations, including as genetic sequencing and protein expression profiling, which can offer useful information about the tumor's traits and possible targets for therapy (Nagrath et al., 2007).

Similarly, technologies for analyzing cfDNA in LOC format allow for non-invasive liquid biopsies, which are less intrusive than conventional tissue biopsies and can be conducted more frequently. These devices employ microfluidic technology to concentrate and identify cfDNA from blood samples, enabling the surveillance of genetic abnormalities and changes linked to cancer (Wan et al., 2017).

Genetic testing

The integration of LOC technologies has also been advantageous for genetic testing, particularly in the context of personalized medicine. LOC devices are capable of conducting a variety of genetic assays, such as gene expression analysis, sequencing, and single-nucleotide polymorphism (SNP) genotyping. These tests are essential for the identification of genetic predispositions to maladies, the guidance of treatment decisions, and the prediction of drug response.

A notable application of LOC devices is in prenatal diagnostics, where they can analyze fetal DNA in maternal blood to identify genetic abnormalities such as Down syndrome, cystic fibrosis, and other inherited health conditions. Park et al. (2013) have demonstrated that this non-invasive method mitigates the risks associated with conventional amniocentesis and chorionic villus sampling.

DNA/RNA AMPLIFICATION

One important use of LOC technology in genetic analysis is the PCR, which is a method for amplifying tiny sections of DNA or RNA. Microfluidic PCR equipment is capable of conducting rapid heat cycling, resulting in a considerable reduction in amplification time compared to traditional approaches. These devices combine the processes of sample preparation, nucleic acid extraction, and amplification into one platform. This makes genetic testing more effective for tasks including detecting pathogens, analyzing genetic mutations, and conducting forensic investigations (Zhang et al., 2006).

Sequencing

Furthermore, LOC devices have made substantial advancements in the field of DNA sequencing. Microfluidic systems have the ability to automate and reduce the size of the intricate processes used in sequencing, including sample preparation, amplification, and detection. This integration not only accelerates the sequencing process but also decreases the necessary sample volume and reagent expenses. Sequencing methods, such as next-generation sequencing (NGS), are being modified to fit into LOC designs. This allows for fast and efficient genetic analysis that is suited for clinical diagnostics and personalized treatment (Mardis, 2008).

Point-of-care testing

The portability and ease of use of LOC devices make them ideal for POC testing, which brings diagnostic capabilities closer to the patient. This is particularly valuable in resource-limited settings where access to centralized laboratories is limited. POC LOC devices can provide rapid diagnostic results in clinics, remote areas, and even in patients' homes, facilitating immediate clinical decision-making and improving patient outcomes.

For example, POC LOC devices have been developed for the rapid diagnosis of diseases such as malaria, tuberculosis, and sexually transmitted infections. These devices often combine microfluidic sample preparation with integrated detection methods, such as immunoassays or nucleic acid amplification, to deliver accurate results quickly and with minimal operator training (Mabey et al., 2004).

Protein assays

Microfluidic LOC devices are employed for a range of protein assays, such as enzyme-linked immunosorbent assays (ELISAs), western blots, and protein-protein interaction studies. These devices have the capability to accurately and precisely identify proteins from small amounts of samples, which makes them highly suitable for the identification and confirmation of biomarkers. LOC platforms for protein assays often include microfluidic channels and chambers that enhance the mixing and reaction kinetics, leading to improved test sensitivity and reduced assay times (Whitesides, 2006).

Cell-based assays

LOC devices have enhanced cell-based experiments by offering a regulated milieu for cell culture, manipulation, and analysis. Microfluidic systems facilitate the examination of cell behavior, pharmacological reactions, and cellular interactions in a tightly regulated environment. LOC devices can be utilized to fabricate organ-on-a-chip (OoC) systems, which replicate the physiological conditions of human organs and are employed for drug testing and disease modeling. These platforms have notable benefits compared to conventional cell culture approaches as they offer settings that are more physiologically appropriate and allow for real-time monitoring of cellular responses (Bhatia & Ingber, 2014).

Multiplexed assays

LOC technologies have the advantage of being able to perform multiplexed tests. Multiplexing enables the concurrent analysis of many analytes from a single sample, hence enhancing the speed and effectiveness of diagnostic procedures. LOC platforms have the ability to include numerous detection zones, each designed to detect a distinct target. This allows for a thorough analysis of biomarkers in a single operation. This feature is especially valuable in fields like cancer diagnostics, where the examination of numerous biomarkers can offer a more precise and thorough diagnosis (Andersson & van den Berg, 2004).

INTEGRATION WITH MOBILE HEALTH TECHNOLOGIES

The integration of LOC devices with mobile health (mHealth) technologies has further expanded their potential in molecular diagnostics. Smartphone-based LOC platforms leverage the processing power, connectivity, and user interface of mobile devices to enhance diagnostic capabilities. These systems can capture, analyze, and transmit data in real-time, providing immediate feedback to healthcare providers and patients.

For instance, smartphone-based LOC devices have been used for the detection of infectious diseases, monitoring chronic conditions, and even for personalized health monitoring. The combination of LOC technologies with mHealth applications facilitates remote diagnostics, telemedicine, and continuous health monitoring, making healthcare more accessible and personalized (Wei et al., 2015).

RECENT INNOVATIONS IN LAB-ON-A-CHIP TECHNOLOGIES

Digital microfluidics

Digital microfluidics (DMF) represents a significant innovation in the field of microfluidics, allowing precise manipulation of discrete droplets on an array of electrodes. Unlike traditional continuous-flow microfluidics, DMF uses electric fields to move, merge, split, and mix droplets. This droplet-based approach offers several advantages, including the ability to handle multiple reagents simultaneously and perform complex assays with high precision and flexibility.

The basic principle of DMF involves applying a voltage to electrodes beneath a hydrophobic dielectric layer, creating an electrostatic force that moves the droplets. This technique can be used to automate and miniaturize various laboratory processes, such as DNA amplification, protein assays, and cell-based assays. The flexibility and reconfigurability of DMF platforms make them particularly suited for applications in personalized medicine and high-throughput screening (Fair, 2007).

One notable application of DMF is in the development of POC diagnostic devices. These devices can perform multiple diagnostic tests on a single platform, providing rapid and accurate results at the patient's bedside. For instance, DMF-based platforms have been developed for the detection of infectious diseases, cancer biomarkers, and genetic mutations. The integration of DMF with other technologies, such as optical detection and mass spectrometry, further enhances the capabilities of these platforms, enabling comprehensive and multiplexed analysis (Abdelgawad & Wheeler, 2009).

Paper-based microfluidics

Paper-based microfluidics is an innovative technique that utilizes the capillary action of paper to propel fluid movement, eliminating the requirement for external pumps. These devices are commonly made by applying hydrophobic barriers on paper substrates, which form specific microchannels that direct the movement of fluids. Due to their simplicity and affordability, paper-based microfluidics are well-suited for use in resource-limited environments and POC applications.

Common methods used in the production of paper-based microfluidic devices include wax printing, inkjet printing, and laser cutting. These techniques enable fast and scalable manufacturing of devices with complex patterns and various testing areas. Paper-based microfluidics is a versatile method for conducting diagnostic tests, such as colorimetric assays, immunoassays, and nucleic acid amplification tests. The findings are commonly interpreted using visual means or with the assistance of a smartphone, rendering these devices highly accessible (Martinez et al., 2010).

A notable benefit of paper-based microfluidics is their capacity to conduct multiplexed assays, allowing for the concurrent identification of many analytes from a solitary sample. This capability is especially advantageous in the diagnosis of intricate disorders, where it is necessary to evaluate many biomarkers in order to achieve a full diagnosis. Moreover, the combination of paper-based microfluidics with smartphone-based imaging and data analysis has created new opportunities for telemedicine and remote diagnostics, improving the availability of healthcare in locations with limited access (Cate et al., 2015).

Organ-on-a-chip technologies

OoC technologies represent a revolutionary approach to modeling human physiology and disease in vitro. These microfluidic devices mimic the structure and function of human organs by combining microfluidic channels with living cells and tissues. OoC platforms provide a more physiologically relevant environment compared to traditional cell culture methods, offering new insights into disease mechanisms, drug responses, and toxicity.

The process of developing OoC devices entails the incorporation of different types of cells, extracellular matrices, and microfluidic channels in order to replicate the microarchitecture and dynamic conditions found in human tissues. Lung-on-a-chip devices replicate the mechanical stretching and air-liquid interface of lung tissue, allowing researchers to investigate respiratory disorders and treatment reactions. Heart-on-a-chip platforms replicate the rhythmic contraction and electrical activity of cardiac tissue, serving as useful models for cardiovascular research (Huh et al., 2010).

OoC technologies have a crucial role in drug development and toxicity testing. These platforms could predict human responses to medications with greater precision compared to animal models, hence decreasing the need for animal testing and expediting the drug development process. OoC devices offer a valuable means of studying disease mechanisms, finding therapeutic targets, and evaluating the effectiveness and safety of new medications by replicating the intricate interactions between many cell types and tissues (Bhatia & Ingber, 2014).

Furthermore, the incorporation of OoC technologies with high-throughput screening and automated data processing facilitates the extensive evaluation of potential drug candidates, hence improving the effectiveness of the drug development process. Utilizing patient-derived cells in OoC platforms presents novel opportunities for personalized medicine, enabling the creation of customized treatments that take into account an individual's unique biological traits (Zhang et al., 2018).

CASE STUDIES: CURRENT CLINICAL TRIALS INVOLVING MICROFLUIDICS TECHNOLOGIES

Rapid detection of infectious diseases

A recent clinical trial focused on the rapid detection of infectious diseases using microfluidic technologies was conducted to address the need for quick and accurate diagnostic tools in outbreak situations. This trial utilized a microfluidic device designed to perform nucleic acid amplification tests (NAATs) for pathogens such as HIV, influenza, and SARS-CoV-2. The device integrated sample preparation, amplification, and detection into a single platform, significantly reducing the time required for diagnosis. Preliminary results demonstrated the device's ability to deliver accurate results within 30 minutes, making it a valuable tool for POC testing in both clinical and field settings (Liu et al., 2021).

Personalized cancer treatment

A current clinical trial is examining the application of microfluidic devices for individualized cancer therapy. This study utilizes a microfluidic chip with the ability to separate CTCs from blood samples. Through the analysis of these cells, researchers seek to acquire knowledge about the genetic composition of cancers and their reaction to various treatments. The microfluidic device enables efficient screening of several medications on isolated CTCs, facilitating the identification of the optimal treatment plan for individual patients. Preliminary results indicate that this method has the potential to greatly influence treatment strategies by customizing medicines for different cancer patients (Nagrath et al., 2007).

Point-of-care testing for cardiovascular diseases

A clinical trial is now being conducted to evaluate the use of a microfluidic device for POC testing of cardiovascular disorders. The device is designed to analyze specific biomarkers that can indicate the presence of heart problems. This gadget utilizes electrochemical detecting techniques to measure the concentrations of cardiac biomarkers, including troponin and BNP (B-type natriuretic peptide), using only a small amount of blood. The trial seeks to assess the device's precision, user-friendliness, and capacity to deliver prompt diagnostic outcomes in emergency scenarios. Preliminary findings have demonstrated potential, since the device produces results that are similar to conventional laboratory testing in just 15 minutes. This emphasizes its ability to accelerate the process of diagnosing and treating acute cardiac conditions (Luppa et al., 2011).

Non-invasive prenatal testing

Microfluidic technologies are making substantial progress in the field of non-invasive prenatal testing (NIPT). Currently, a clinical trial is evaluating a microfluidic device that is specifically designed to separate and examine cell-free fetal DNA (cffDNA) from blood samples taken from pregnant women. This device is designed to identify genetic abnormalities, including Down syndrome, cystic fibrosis, and other hereditary disorders. The trial focuses on evaluating the precision, security, and effectiveness of microfluidic technology in comparison to traditional approaches. Initial evidence suggests that the microfluidic device can yield reliable results with a small amount of sample, hence decreasing the likelihood of complications and discomfort typically associated with conventional invasive procedures (Bianchi et al., 2014).

Sepsis diagnosis and monitoring

A clinical trial investigating the rapid diagnosis and monitoring of sepsis is utilizing a microfluidic device to detect sepsis biomarkers in blood samples. Sepsis is a life-threatening condition that requires immediate diagnosis and intervention. The microfluidic device under trial integrates immunoassay techniques to detect biomarkers such as procalcitonin and IL-6 (interleukin-6), providing results within an hour. The trial aims to evaluate the device's ability to quickly and accurately diagnose sepsis, potentially improving patient outcomes through earlier and more targeted treatments. Initial results are promising, demonstrating the device's capability to deliver rapid diagnostics in critical care environments (Kumar et al., 2011).

CONCLUSION

Over the past decade, there has been substantial progress in microfluidics and LOC technologies. These advancements have brought about an advancement in many different fields like medical diagnostics, environmental monitoring, and chemical synthesis. Precise manipulation of small fluid volumes in microchannels allows for accurate control of fluid dynamics, which is essential for applications that demand regulated environments. The development of microfabrication techniques such as photolithography, soft lithography, and 3D printing has made it easier to create intricate microfluidic networks with great accuracy. The advancements in materials science have resulted in the development of polymers that are both biocompatible and chemically resistant, hence broadening the range of applications for these devices. The combination of biosensors with microfluidics has led to the development of extremely sensitive diagnostic platforms that are well-suited for continuous monitoring. In addition, advancements in DMF, paper-based microfluidics, and OoC technologies have expanded the possibilities in POC testing, personalized medicine, and drug development. These technologies are currently undergoing validation in clinical trials for several applications, including quick detection of infectious diseases, customized treatment for cancer, diagnostics for cardiovascular conditions, non-invasive testing for prenatal health, and monitoring of sepsis. The ongoing development of microfluidics holds the potential to significantly improve the functionality and availability of diagnostic tools, which will eventually contribute to better patient outcomes and increased efficiency in healthcare.

REFERENCES

Abdelgawad, M., & Wheeler, A. R. (2009). The digital microfluidic toolbox: A review. Advanced Materials, 21(8), 920–925. https://doi.org/10.1002/adma.200801186

Andersson, H., & van den Berg, A. (2004). Microtechnologies and nanotechnologies for single-cell analysis. Current Opinion in Biotechnology, 15(1), 44–49. https://doi.org/10.1016/j.copbio.2004.02.004

Becker, H., & Gärtner, C. (2008). Polymer microfabrication technologies for microfluidic systems. Analytical and Bioanalytical Chemistry, 390(1), 89–111. https://doi.org/10.1007/s00216-007-1692-2

Bhatia, S. N., & Ingber, D. E. (2014). Microfluidic organs-on-chips. Nature Biotechnology, 32(8), 760–772. https://doi.org/10.1038/nbt.2989

Bhattacharyya, A., & Klapperich, C. M. (2005). Thermoplastic microfluidic device for on-chip purification of nucleic acids for disposable diagnostics. Analytical Chemistry, 77(14), 4293–4299. https://doi.org/10.1021/ac051449j

Bianchi, D. W., Chiu, R. W. K., & Lo, Y. M. D. (2014). Noninvasive prenatal testing: The paradigm is shifting rapidly. Prenatal Diagnosis, 34(6), 603–611. https://doi.org/10.1002/pd.4355

Cate, D. M., Adkins, J. A., Mettakoonpitak, J., & Henry, C. S. (2015). Recent developments in paper-based microfluidic devices. Analytical Chemistry, 87(1), 19–41. https://doi.org/10.1021/ac503968p

Chin, C. D., Linder, V., & Sia, S. K. (2011). Commercialization of microfluidic point-of-care diagnostic devices. Lab on a Chip, 12(12), 2114–2134. https://doi.org/10.1039/C2LC21204H

Dittrich, P. S., & Manz, A. (2006). Lab-on-a-chip: Microfluidics in drug discovery. Nature Reviews Drug Discovery, 5(3), 210–218. https://doi.org/10.1038/nrd1985

Fair, R. B. (2007). Digital microfluidics: Is a true lab-on-a-chip possible? Microfluidics and Nanofluidics, 3(3), 245–281. https://doi.org/10.1007/s10404-007-0161-8

Fiorini, G. S., & Chiu, D. T. (2005). Disposable microfluidic devices: Fabrication, function, and application. BioTechniques, 38(3), 429–446. https://doi.org/10.2144/05383RV01

Gervais, L., de Rooij, N., & Delamarche, E. (2011). Microfluidic chips for point-of-care immunodiagnostics. Advanced Materials, 23(24), H151–H176. https://doi.org/10.1002/adma.201100464

Huh, D., Matthews, B. D., Mammoto, A., Montoya-Zavala, M., Hsin, H. Y., & Ingber, D. E. (2010). Reconstituting organ-level lung functions on a chip. Science, 328(5986), 1662–1668. https://doi.org/10.1126/science.1188302

Kumar, A., Roberts, D., Wood, K. E., Light, B., Parrillo, J. E., Sharma, S., ... Cheang, M. (2011). Duration of hypotension before initiation of effective antimicrobial therapy is the critical determinant of survival in human septic shock. Critical Care Medicine, 34(6), 1589–1596. https://doi.org/10.1097/01.CCM.0000217961.75225.E9

Levy, S., Wesner, D. A., & Osten, W. (2013). Fabrication of microfluidic devices by micro-milling and laser ablation. Journal of Micromechanics and Microengineering, 23(6), 065009. https://doi.org/10.1088/0960-1317/23/6/065009

Liu, Y., Liu, Q., Wu, J., Zhang, X., & Wang, Z. (2021). Current advances in COVID-19 diagnostic devices. Journal of Pharmaceutical Analysis, 11(2), 136–146. https://doi.org/10.1016/j.jpha.2020.07.003

Luppa, P. B., Müller, C., Schlichtiger, A., & Schlebusch, H. (2011). Point-of-care testing (POCT): Current techniques and future perspectives. Trends in Analytical Chemistry, 30(6), 887–898. https://doi.org/10.1016/j.trac.2011.01.019

Mabey, D., Peeling, R. W., Ustianowski, A., & Perkins, M. D. (2004). Diagnostics for the developing world. Nature Reviews Microbiology, 2(3), 231–240. https://doi.org/10.1038/nrmicro841

Madou, M. J. (2002). Fundamentals of microfabrication: The science of miniaturization. CRC Press.

Mardis, E. R. (2008). Next-generation DNA sequencing methods. Annual Review of Genomics and Human Genetics, 9, 387–402. https://doi.org/10.1146/annurev.genom.9.081307.164359

Martinez, A. W., Phillips, S. T., Whitesides, G. M., & Carrilho, E. (2010). Diagnostics for the developing world: Microfluidic paper-based analytical devices. Analytical Chemistry, 82(1), 3–10. https://doi.org/10.1021/ac9013989

McDonald, J. C., Duffy, D. C., Anderson, J. R., Chiu, D. T., Wu, H., Schueller, O. J., & Whitesides, G. M. (2000). Fabrication of microfluidic systems in poly(dimethylsiloxane). Electrophoresis, 21(1), 27–40. https://doi.org/10.1002/(SICI)1522-2683(20000101)21:1<27::AID-ELPS27>3.0.CO;2-C

Nagrath, S., Sequist, L. V., Maheswaran, S., Bell, D. W., Irimia, D., Ulkus, L., ... Toner, M. (2007). Isolation of rare circulating tumour cells in cancer patients by microchip technology. Nature, 450(7173), 1235–1239. https://doi.org/10.1038/nature06385

Nguyen, N. T., Shaegh, S. A. M., Kashaninejad, N., & Phan, D. T. (2018). Design, fabrication and characterization of droplet-based microfluidic systems. Lab on a Chip, 18(1), 8–27. https://doi.org/10.1016/j.addr.2013.05.008

Park, J. M., Kim, S. M., & Park, D. (2013). Microfluidic systems for advanced diagnostics: A new toolbox for multiplexed and integrated assays. Lab on a Chip, 13(3), 509–513.

Sackmann, E. K., Fulton, A. L., & Beebe, D. J. (2014). The present and future role of microfluidics in biomedical research. Nature, 507(7491), 181–189. https://doi.org/10.1038/nature13118

Squires, T. M., & Quake, S. R. (2005). Microfluidics: Fluid physics at the nanoliter scale. Reviews of Modern Physics, 77(3), 977–1026. https://doi.org/10.1103/RevModPhys.77.977

Stone, H. A., & Kim, S. (2001). Microfluidics: Basic issues, applications, and challenges. AIChE Journal, 47(6), 1250–1254. https://doi.org/10.1002/aic.690470602

Teh, S.-Y., Lin, R., Hung, L.-H., & Lee, A. P. (2008). Droplet microfluidics. Lab on a Chip, 8(2), 198–220. https://doi.org/10.1039/B715524G

Unger, M. A., Chou, H. P., Thorsen, T., Scherer, A., & Quake, S. R. (2000). Monolithic microfabricated valves and pumps by multilayer soft lithography. Science, 288(5463), 113–116. https://doi.org/10.1126/science.288.5463.113

Wan, J. C. M., Massie, C., Garcia-Corbacho, J., Mouliere, F., Brenton, J. D., Caldas, C., ... Rosenfeld, N. (2017). Liquid biopsies come of age: Towards implementation of circulating tumour DNA. Nature Reviews Cancer, 17(4), 223–238. https://doi.org/10.1038/nrc.2017.7

Wei, Q., Nagi, R., Sadeghi, K., Feng, S., Yan, E., Ki, S. J., ... Ozcan, A. (2015). Detection and spatial mapping of mercury contamination in water samples using a smart-phone. ACS Nano, 8(2), 1121–1129.

Whitesides, G. (2006). The origins and the future of microfluidics. Nature 442, 368–373. https://doi.org/10.1038/nature05058

Whitesides, G. M., & Stroock, A. D. (2001). Flexible methods for microfluidics. Physics Today, 54 (6), 42–48.
Xia, Y., & Whitesides, G. M. (1998). Soft lithography. Annual Review of Materials Science, 28(1), 153–184.
Yetisen, A. K., Akram, M. S., & Lowe, C. R. (2013). Paper-based microfluidic point-of-care diagnostic devices. Lab on a Chip, 13(12), 2210–2251. https://doi.org/10.1039/C3LC50169H
Zhang, B., Korolj, A., Lai, B. F., & Radisic, M. (2018). Advances in organ-on-a-chip engineering. Nature Reviews Materials, 3(8), 257–278.
Zhang, C., Xu, J., Ma, W., & Zheng, W. (2006a). PCR microfluidic devices for DNA amplification. Biotechnology Advances, 24(3), 243–284.
Zhang, C., Zhao, Z., Abdul Rahim, N. A., van Noort, D., & Yu, H. (2006b). Towards a human-on-chip: Culturing multiple cell types on a chip with compartmentalized microenvironments. Lab on a Chip, 9(22), 3185–3192. https://doi.org/10.1039/B915147H
Zheng, X., & Yang, Z. (2023). Recent advances in electrochemical biosensors for the detection of foodborne pathogens: Current perspective and challenges. Foods, 12(14), 2795.

Chapter 11

Transcriptomics analysis in molecular diagnostics

Amenah S. Al-otaibi

Department of Biology, Biodiversity Genomics Unit,
Faculty of Science, University of Tabuk, Tabuk, Saudi Arabia

Marya Ahsan

Department of Pharmacology, College of Medicine, Imam Mohammad
Ibn Saud Islamic University (IMSIU), Riyadh, Saudi Arabia

*Ayaz Khurram Mallick**

Department of Clinical Biochemistry, College of Medicine,
King Khalid University, Abha, Saudi Arabia

PROTEOMICS INTEGRATION

Transcriptomics has greatly enhanced our capacity to comprehend and identify diseases at the molecular level. Transcriptomics offers a thorough understanding of gene expression alterations linked to different diseases by analyzing a complete set of RNA transcripts generated by the genome. Having an extensive understanding of gene expression patterns and regulatory processes is essential for the identification of new therapeutic targets and the development of efficient treatments.

Transcriptomics offers extensive knowledge about gene expression patterns, uncovering crucial details about the processes underlying diseases. Through the comparison of the transcriptomes of healthy and diseased tissues, researchers can detect genes that exhibit differential expression and reveal previously unknown pathways that play a role in the development of diseases (Wang et al., 2014). Transcriptomics has had a significant impact on cancer research by contributing to the identification of different forms of cancer, the comprehension of tumor heterogeneity, and the discovery of possible targets for therapy (Cancer Genome Atlas Research Network, 2014).

RNA-Seq has emerged as a potent technique for identifying and confirming biomarkers for a range of diseases. Transcriptomic analysis can detect RNA-based biomarkers that indicate the presence of a disease, its course, and the response to treatment. Non-coding RNAs, including microRNAs and long non-coding RNAs, have been shown to have important functions in regulating diseases and are being more recognized as potential biomarkers (Esteller, 2011). The utilization of these biomarkers has the potential to greatly improve the early detection and personalized treatment approaches (Fang et al., 2018).

Integrating transcriptomics into clinical practice enables personalized medicine by customizing therapies according to individual gene expression profiles. RNA-Seq data has the ability to predict how patients will respond to particular medicines, allowing for the identification of the most efficient treatment plans (Schwartz et al., 2016). Transcriptomics-guided

* Corresponding author.

DOI: 10.1201/9781003489177-11

medicines have effectively been utilized in the field of oncology, leading to improved patient outcomes (Roden et al., 2017).

Transcriptomic analysis has played a crucial role in the identification of novel therapeutic targets. Through the process of clarifying the functions of different genes and non-coding RNAs in diseases, researchers can create novel drugs that specifically focus on these molecules. The discovery of distinct RNA splicing variations and their correlation with diseases has created new opportunities for therapeutic intervention (Lee & Rio, 2015).

Transcriptomics expedites therapeutic intervention by offering comprehensive insights into alterations in gene expression in response to drugs. This allows for the identification of potential off-target effects and the mechanisms of drug resistance, ultimately leading to the development of safer and more effective therapies (Hodos et al., 2016).

Single-cell RNA-Seq has enhanced the potential of transcriptomics in molecular diagnostics. Single-cell analyses provide a high-resolution view of cellular heterogeneity, revealing rare cell populations and transient cell states that are crucial for understanding disease progression and treatment response (Papalexi & Satija, 2018). These findings have significant implications within the fields of cancer research, developmental biology, and immunology.

However, transcriptomics, despite its ability to bring about significant changes, encounters multiple challenges such as intricate data, high costs, and the requirement for standardized procedures. Nevertheless, continuous progress in sequencing technology and bioinformatics has addressed these challenges, hence enhancing the accessibility and dependability of transcriptomics for clinical applications (Conesa et al., 2016).

METHODOLOGICAL APPROACHES

High-throughput sequencing technologies have revolutionized the field of transcriptomics by enabling the detailed and comprehensive analysis of the transcriptome. These technologies provide a powerful means to study gene expression, identify novel transcripts, and explore the complexity of RNA splicing and regulation. Among the high-throughput sequencing technologies, RNA sequencing (RNA-Seq) stands out as the gold standard for transcriptome analysis due to its versatility, sensitivity, and accuracy.

Basics of RNA-Seq

RNA sequencing (RNA-Seq) is a next-generation sequencing (NGS) technology used to study the transcriptome, the complete set of RNA transcripts produced by the genome under specific circumstances. RNA-Seq allows for the quantification of gene expression levels, the identification of novel transcripts, and the detection of alternative splicing events. As shown in **Figure 11.1**, the technology works by converting RNA molecules into complementary DNA (cDNA), which is then sequenced to produce millions of short reads that represent the RNA present in the sample (Wang et al., 2009).

Basic steps of RNA-Seq

RNA Isolation: The first step involves isolating high-quality RNA from the sample. The integrity and purity of RNA are essential for the successful completion of the subsequent procedures. TRIzol extraction and column-based purification kits are frequently employed techniques for RNA extraction. The quality of the extracted RNA is typically assessed using

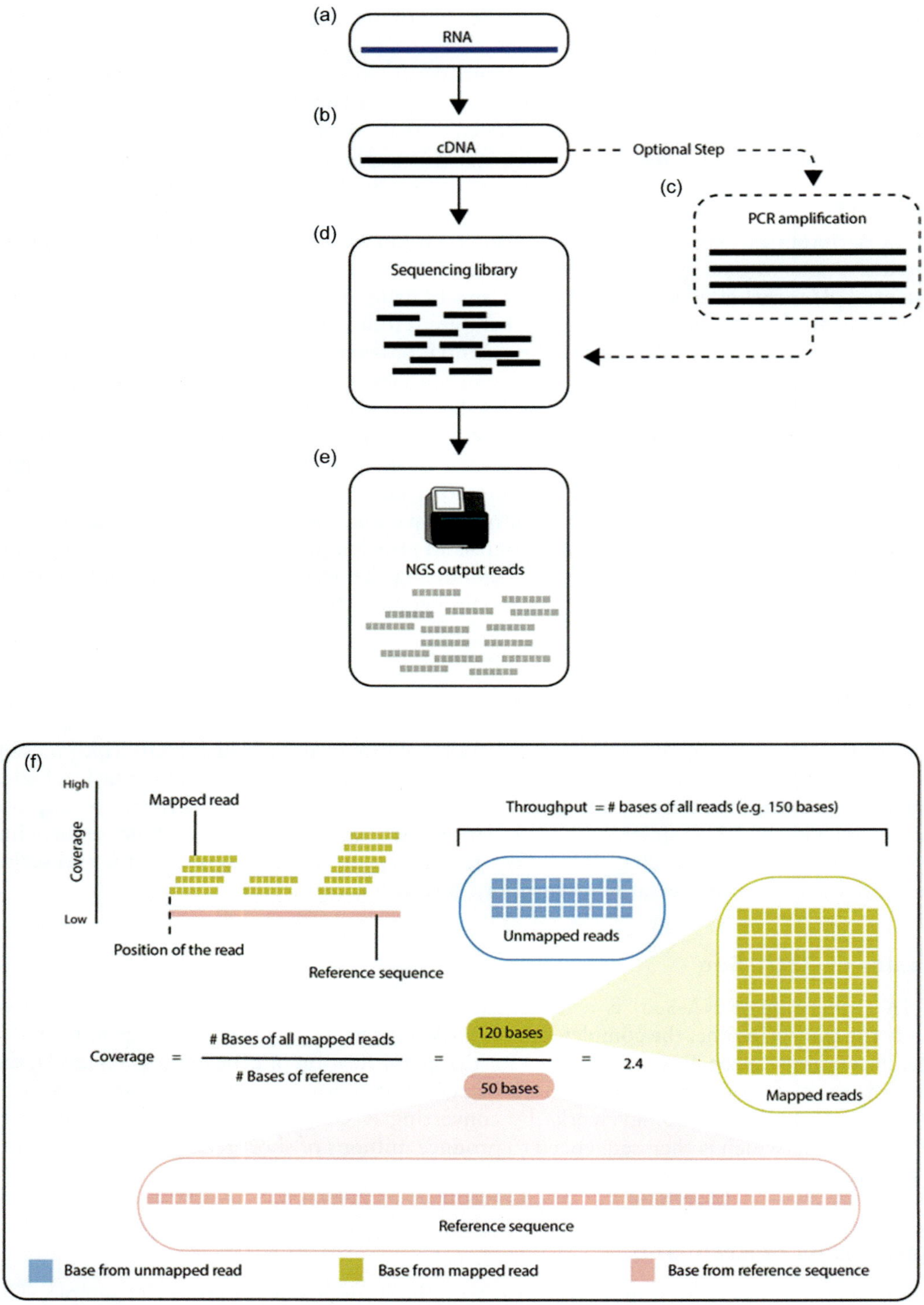

Figure 11.1 Workflow of RNA sequencing via NGS, (a) including RNA isolation, (b) cDNA synthesis, (c) PCR amplification, (d) library preparation, sequencing, and (e) data interpretation using (f) coverage and throughput metrics.

instruments such as the Agilent Bioanalyzer, which provides an RNA Integrity Number (RIN) score.

Library Preparation: The isolated RNA is converted into a cDNA library through a series of steps. First, the RNA is fragmented into smaller pieces. Then, reverse transcription is performed to convert the RNA fragments into cDNA. Sequencing adapters are added to both ends of the cDNA fragments to facilitate sequencing. Depending on the specific use case, different approaches can be utilized, such as employing poly(A) selection to concentrate messenger RNA (mRNA) or using ribosomal RNA (rRNA) depletion to decrease the amount of rRNA.

Sequencing: Subsequently, the cDNA library that has been generated is subjected to sequencing utilizing a high-throughput sequencing platform. During sequencing, each cDNA fragment is read multiple times to generate short sequences known as reads. Typically, these reads have a length of 50–150 base pairs, while different platforms might produce larger reads. The sequencing depth, defined as the quantity of reads produced for each transcript, is a crucial determinant of the sensitivity and precision of RNA-Seq analysis.

Data Analysis: The raw sequencing reads are processed and analyzed using bioinformatics tools. This involves several steps, including quality control, alignment of reads to a reference genome or transcriptome, quantification of gene expression levels, and identification of differentially expressed genes and splice variants. Advanced analyses, such as transcript assembly, fusion gene detection, and single-cell RNA-Seq, can also be performed depending on the research objectives.

Platforms for RNA-Seq

RNA-Seq is routinely performed on several high-throughput sequencing systems, each with special qualities and benefits. Platform choice can rely on elements including cost, accuracy, read length, and throughput. Among the most often utilized systems are Oxford Nanopore Technologies (ONTs), Illumina, and Pacific Biosciences (PacBio)

Illumina Sequencing: The HiSeq and NovaSeq series are widely utilized for RNA-Seq due to their superior capacity, precision, and cost-efficiency. Illumina sequencing generates short reads, usually ranging from 50 to 150 base pairs in size, that are suitable for gene expression profiling and alternative splicing studies. The platform's high throughput enables it to generate a significant number of reads in each run, making it well-suited for extensive transcriptome investigations and clinical applications (Bentley et al., 2008).

Pacific Biosciences (PacBio): PacBio technologies, such as the Sequel system, provide the ability to sequence long reads. This is beneficial for analyzing intricate transcript structures, identifying whole isoforms, and investigating repetitive areas. PacBio's Single Molecule Real-Time (SMRT) sequencing technique offers readings with exceptional accuracy, albeit at a greater expense in comparison to Illumina. Long-read sequencing is extremely useful for de novo transcriptome assembly and for studying the diversity of isoforms (Eid et al., 2009).

ONTs: Ultra-long reads and the capacity to sequence natural RNA molecules directly without the need for reverse transcription are provided by ONT platforms like the MinION and PromethION. Studying RNA alterations and obtaining full-length transcripts in a single read are two applications of this technology that are very helpful. Because ONT platforms are flexible and portable, they can be used for both clinical and field-based applications. ONT's primary benefit is its capacity to generate data in real-time, which can be essential for time-sensitive diagnostic applications (Jain et al., 2016).

Applications of RNA-Seq in molecular diagnostics

RNA-Seq has a wide range of applications in molecular diagnostics and biomedical research. Some notable applications include:

Gene Expression Profiling: RNA-Seq is employed to measure the levels of gene expression in various conditions, tissues, or cell types. This helps in the identification of genes that exhibit differential expression in diseases, such as cancer, hence offering valuable insights into disease processes and prospective targets for therapeutic intervention (Wang et al., 2014).

Alternative Splicing Analysis: RNA-Seq allows for the identification of alternative splicing events, which can result in the generation of several transcript isoforms from a single gene. Abnormal splicing is linked to a range of diseases, and RNA-Seq offers a thorough understanding of the splicing patterns (Katz et al., 2010).

Single-Cell Transcriptomics: Single-cell RNA-Seq enables the examination of gene expression at the level of individual cells, revealing the diversity of cells within tissues and detecting rare cell types. This is particularly valuable in cancer research, immunology, and developmental biology (Papalexi & Satija, 2018).

Novel Transcript Discovery: RNA-Seq has the ability to detect new transcripts, such as genes and non-coding RNAs, that have not been previously identified. This helps to get a deeper knowledge of the transcriptome and its regulatory functions (Guttman et al., 2009).

Clinical Diagnostics and Precision Medicine: RNA-Seq is being used more and more in clinical diagnostics to discover biomarkers for disease diagnosis, prognosis, and therapy response. Additionally, it has a vital function in precision medicine by tailoring therapies according to individual gene expression patterns (Schwartz et al., 2016).

MICROARRAY-BASED METHODS

Microarrays, commonly referred to as DNA chips or gene chips, are solid surfaces that contain a multitude of fixed DNA probes that match known genes (**Figure 11.2**). These probes have the ability to bind with DNA or RNA sequences that are complementary in a given sample, enabling the assessment of gene expression levels for thousands of genes at the same time.

Basics of microarray technology

Microarray technology relies on the concept of hybridization, in which complementary nucleic acid strands bind to one another. The fundamental stages of a microarray experiment are sample preparation, hybridization, washing, and detection. The method enables the comparative investigation of gene expression under various situations, making it a potent tool for identifying differentially expressed genes.

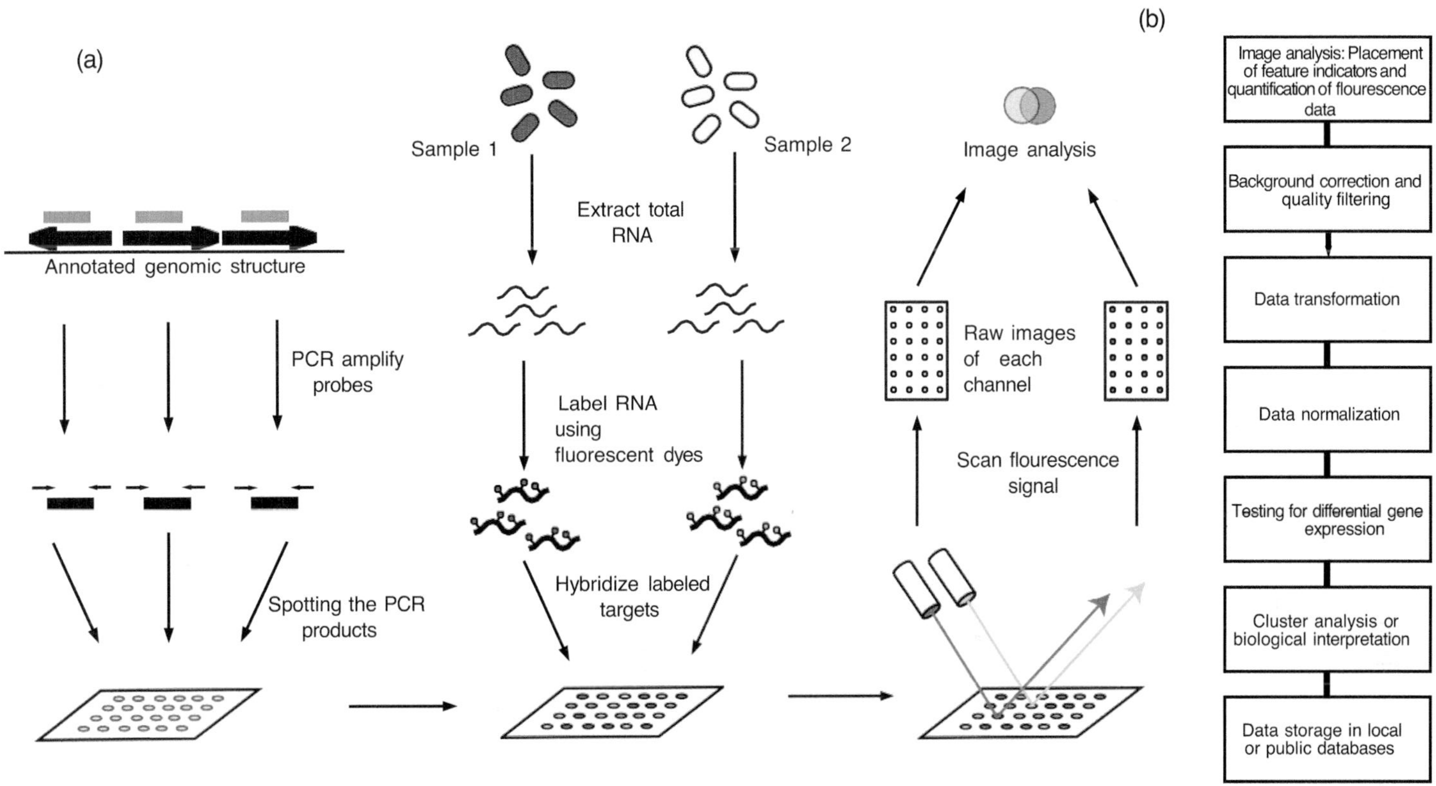

Figure 11.2 Microarray-based gene expression analysis showing (a) RNA labeling, hybridization to probes, scanning, and (b) computational analysis.

THE KEY STEPS IN A MICROARRAY EXPERIMENT ARE AS FOLLOWS:

Sample Preparation: The process involves extracting the whole of the RNA from the biological samples. Subsequently, the RNA is transformed into either complementary DNA (cDNA) or complementary RNA (cRNA), which is subsequently marked with fluorescent dyes. Typical dyes used are Cy3 (green) and Cy5 (red), enabling the simultaneous comparison of two samples on a single microarray.

Hybridization: The labeled cDNA or cRNA is hybridized to a microarray, where it binds to the complementary DNA probes on the array. Hybridization is typically performed under controlled conditions to ensure specific binding.

Washing: Unbound and nonspecifically bound nucleic acids are washed away to reduce background noise. The stringency of the wash conditions can be adjusted to optimize the specificity and sensitivity of the hybridization.

Detection: The microarray is scanned using a laser scanner that detects the fluorescent signals emitted by the labeled probes. The intensity of the fluorescence at each spot on the array corresponds to the abundance of the complementary RNA in the sample.

Data Analysis: The raw fluorescence data are processed and normalized to account for technical variations. The normalized data are then analyzed to identify differentially expressed genes and to perform functional annotation and pathway analysis.

Types of microarrays

Several types of microarrays are used in transcriptomics, each with specific applications and advantages:

cDNA Microarrays: The original version of microarrays consisted of cDNA fragments obtained from well-known genes and spotted onto a glass plate. cDNA microarrays are commonly employed for conducting comparative gene expression analyses.

Oligonucleotide Microarrays: These microarrays use short synthetic DNA oligonucleotides (typically 25–60 nucleotides in length) as probes. The probes are designed to be highly specific to their target sequences, allowing for greater sensitivity and specificity. Affymetrix GeneChips are a well-known example of oligonucleotide microarrays.

Tiling Microarrays: Tiling microarrays cover large regions of the genome with overlapping probes, allowing for the detection of novel transcripts, alternative splicing events, and non-coding RNAs. They are particularly useful for exploring unannotated regions of the genome.

Exon Microarrays: These microarrays contain probes targeting individual exons of genes, enabling the study of alternative splicing and exon-level expression changes. They are useful for identifying splice variants and understanding the complexity of gene regulation.

APPLICATIONS OF MICROARRAY-BASED METHODS

Despite the advent of RNA-Seq, microarrays continue to be widely used in various applications due to their cost-effectiveness, ease of use, and established protocols. Key applications include:

Gene Expression Profiling: Microarrays are commonly employed to compare gene expression levels across different situations, including healthy versus sick tissues, treated versus untreated samples, or different developmental stages. This application has played a crucial role in finding genes linked to illnesses, comprehending disease processes, and uncovering possible targets for therapy.

Disease Classification and Diagnosis: Microarrays have been employed to classify diseases according to their gene expression patterns. In the field of cancer research, scientists have created microarray-based gene expression profiles to differentiate between various types of cancers, predict patient prognosis, and assist in making decisions about treatment (van 't Veer et al., 2002).

Pharmacogenomics: Microarrays are employed to investigate the impact of genetic differences on individual responses to drugs. Through the examination of gene expression patterns, researchers are able to discover biomarkers that might predict the effectiveness and toxicity of drugs. This knowledge is crucial for the advancement of personalized medicine (Wang et al., 2011).

Toxicogenomics: This application entails investigating the impact of toxic substances on gene expression. Microarrays are utilized to detect alterations in gene expression caused by environmental toxins, pharmaceuticals, and other substances. This enables the identification of underlying processes of toxicity and the discovery of possible biomarkers for exposure (Hamadeh et al., 2002).

Comparative Genomics: Microarrays are a useful tool for comparing gene expression patterns among various species, strains, or situations. This helps understand evolutionary links and determine if genes have similar functions across different organisms.

Transcriptomics is expected to advance by combining microarray and RNA-Seq technologies. Hybrid techniques can exploit the advantages of both technologies by employing microarrays for initial high-throughput screening and RNA-Seq for more thorough and extensive analysis. Furthermore, progress in microarray technology continues in enhancing its efficiency and versatility. Microarrays are being improved by innovations like high-density arrays, enhanced probe design, and more advanced data analysis methods. These improvements are increasing the sensitivity and specificity of microarrays.

APPLICATIONS OF TRANSCRIPTOMICS: DETECTING ABERRANT TRANSCRIPTION IN HUMAN DISEASE

Transcriptomics plays a crucial role in identifying abnormal gene expression, which is vital for understanding the molecular mechanisms underlying illnesses and improving clinical diagnostics and treatment approaches. Transcriptomics offers comprehensive insight into mRNA expression profile, gene fusions, and alternative transcripts, which are especially significant in the context of cancer and other complex disorders. This section explores the importance of analyzing mRNA expression profiles and identifying gene fusions and alternative transcripts in human diseases.

OVERVIEW OF mRNA EXPRESSION PROFILING

mRNA expression profiling is the process of quantifying the abundance of messenger RNA (mRNA) molecules in a sample in order to identify patterns of gene expression. Profiling is crucial for the identification of genes that are expressed differently, the understanding of disease mechanisms, and the development of diagnostic and therapeutic approaches. Transcriptomics, particularly with the use of modern techniques like as RNA sequencing (and microarrays, has greatly enhanced our capacity to perform thorough analysis of mRNA expression patterns.

ROLE IN CANCER DIAGNOSTICS

Cancer is defined by atypical gene expression patterns caused by genetic mutations, epigenetic alterations, and disrupted signaling pathways. mRNA expression profiling is very crucial in cancer diagnoses for several reasons:

Identification of Cancer Subtypes: Various forms of cancer, including specific variations within a particular kind, have unique patterns of gene expression. mRNA expression profiling aids in categorizing cancers according to their molecular characteristics. For instance, breast cancer may be categorized into subtypes, including luminal A, luminal B, HER2-enriched, and triple-negative, based on their gene expression patterns (Perou et al., 2000). The molecular categorization is essential for choosing suitable treatment options.

Prognostic and Predictive Biomarkers: Gene expression characteristics determined by mRNA profiling can function as prognostic and predictive biomarkers. Prognostic biomarkers offer insights into the probable progression of a disease, whereas predictive biomarkers show the anticipated reaction to a particular therapy. The MammaPrint test employs a 70-gene expression profile to forecast the likelihood of breast cancer recurrence and provide guidance for treatment choices (van 't Veer et al., 2002).

Monitoring Treatment Response: mRNA expression profiling can be used to monitor the response to therapy and detect minimal residual disease. Changes in the expression levels of specific genes can indicate whether a treatment is effective or if the cancer is likely to relapse. This approach allows for the early detection of treatment failure and the adjustment of therapeutic strategies.

Understanding Tumor Microenvironment: mRNA expression analysis offers valuable insights into the interactions that occur between cancer cells and their microenvironment. This encompasses the examination of immune cells, stromal cells, and signaling molecules that have a role in the development and spread of tumors. Gaining a comprehensive understanding of these interactions is crucial for the development of precise and focused treatments and immunotherapies.

GENE FUSIONS AND ALTERNATIVE TRANSCRIPTS

Gene fusions and alternative splicing events are important components of transcriptional control that can contribute to the development of disease. Gene fusions arise from chromosomal rearrangements, causing the fusion of two distinct genes. This leads to the formation of a hybrid protein that exhibits modified functionality. Alternative splicing is the mechanism by which various combinations of exons are fused together to generate several transcript variants from a solitary gene.

ROLE IN DISEASE DIAGNOSTICS AND THERAPY

Gene fusions and alternative splicing events play significant roles in the pathogenesis of various diseases, particularly cancer. Their identification has important diagnostic and therapeutic implications:

Gene Fusions: Gene fusions can function as oncogenic drivers, facilitating unregulated cell growth and survival. Detecting these fusions is essential for accurately detecting particular forms of cancer and formulating precise therapeutic approaches. An illustration of this is the BCR-ABL fusion gene, which occurs as a result of the translocation between chromosomes 9 and 22. This gene is a distinctive characteristic of chronic myeloid leukemia (CML). The discovery of BCR-ABL has resulted in the creation of targeted treatments like imatinib, which selectively hinders the function of the fusion protein (Druker et al., 2001).

Alternative Splicing: Abnormal splicing events can generate isoforms that have different or even opposing activities, which can contribute to the advancement of diseases. Alternative splicing is a mechanism that produces different versions of proteins and controls gene expression. However, when this process is not properly regulated, it can lead to the development of different types of cancers and other disorders. For instance, the CD44 gene undergoes alternative splicing, resulting in the production of several isoforms that have a role in the migration, invasion, and spread of cancer cells (Brown et al., 2001).

CLINICAL EXAMPLES AND STUDIES

The identification of abnormal gene expression caused by gene fusions and alternative splicing events is crucial for the diagnosis and treatment of many types of malignancies and other disorders. Multiple clinical investigations have emphasized the significance of finding these deviations in gene expression, which has resulted in the creation of specific treatments and enhanced patient outcomes.

FUSION GENES IN CANCER

Fusion genes result from chromosomal rearrangements that join parts of two different genes, leading to the production of hybrid proteins with oncogenic potential. The identification of these fusion genes has revolutionized cancer diagnostics and therapy.

BCR-ABL fusion gene in chronic myeloid leukemia

The BCR-ABL fusion gene, which occurs due to a translocation between chromosomes 9 and 22 (known as the Philadelphia chromosome), is a characteristic feature of CML. The fusion gene contains the genetic information for a tyrosine kinase that is always active and promotes uncontrolled cell growth (Rowley, 1973). The identification of BCR-ABL prompted the creation of imatinib, a tyrosine kinase inhibitor that selectively focuses on the fusion protein. Imatinib has significantly improved the prognosis of patients with CML, changing it from a terminal illness to a controllable long-term condition (Druker et al., 2001).

EML4-ALK fusion gene in non-small cell lung cancer

The EML4-ALK fusion gene is found in a specific group of individuals with non-small cell lung cancer (NSCLC). It occurs due to a chromosomal inversion on chromosome 2, which causes the EML4 and ALK genes to fuse together. The fusion of ALK kinase results in its continuous activation, which in turn stimulates the development of tumors (Soda et al., 2007). The identification of EML4-ALK fusion has important clinical implications, since patients with this fusion can be effectively treated with targeted treatments like crizotinib. Crizotinib hampers the action of ALK kinase, resulting in the regression of tumors and extended life in patients diagnosed with ALK-positive NSCLC (Kwak et al., 2010).

TMPRSS2-ERG fusion gene in prostate cancer

The TMPRSS2-ERG fusion gene, present in over 50% of prostate malignancies, is formed by a translocation event that combines the androgen-regulated TMPRSS2 promoter with the ERG oncogene. The fusion of genes results in the excessive production of ERG, which promotes the advancement of prostate cancer (Tomlins et al., 2005). The discovery of TMPRSS2-ERG has yielded important information on the molecular processes underlying prostate cancer and has significant implications for targeted treatment and prognosis.

ALTERNATIVE SPLICING VARIANTS IN CANCER

Alternative splicing enables the generation of several transcript isoforms from a single gene, which enhances protein diversity and thus regulates gene expression. Abnormal splicing events can generate isoforms with different or even opposing functions, contributing to the genesis and advancement of cancer.

CD44 alternative splicing in breast cancer

The CD44 gene undergoes significant alternative splicing to generate several isoforms, such as CD44 standard (CD44s) and variant isoforms (CD44v). These isoforms have distinct functions in cell adhesion, migration, and signaling. The presence of certain CD44 isoforms is linked to the advancement and spread of breast cancer. CD44v6 is correlated with increased metastatic capacity, whereas CD44s is connected to the characteristics of cancer stem cells (Brown et al., 2001). Therapeutic approaches in breast cancer show potential when focusing on CD44 splicing variations.

MDM2 splicing variants in cancer

The MDM2 gene, responsible for encoding a protein that inhibits the tumor suppressor p53, undergoes alternative splicing to generate multiple forms of the protein. Certain MDM2 isoforms do not possess the p53-binding domain, which leads to increased oncogenic capability and resistance to chemotherapy. Poor prognosis in several malignancies, such as sarcomas and breast cancer, is linked to the presence of particular MDM2 splicing variants (Bartel et al., 2002). Gaining insight into the control of MDM2 splicing presents possibilities for creating targeted therapeutics that modulate splicing events.

FGFR3 splicing variants in bladder cancer

The FGFR3 gene undergoes frequent mutations and alternative splicing in bladder cancer. The FGFR3b and FGFR3c isoforms, generated by alternative splicing, have distinct preferences for binding ligands and varying levels of oncogenicity. FGFR3c is linked to heightened tumor aggressiveness and poor prognosis (Sibley et al., 2001). Targeting FGFR3 alternative splicing isoforms and their subsequent signaling cascades provides a promising therapeutic approach for bladder cancer.

CHALLENGES AND FUTURE DIRECTIONS

Although there have been significant improvements in the detection of gene fusions and alternative splicing events, there are still some obstacles that need to be overcome. These factors encompass the intricacy of splicing patterns, the requirement for increased sensitivity and specificity, and the analysis of clinical significance. Continuing research and advancements in technology are focused on tackling these problems and improving the usefulness of transcriptomics in clinical diagnosis.

Improving detection sensitivity and specificity

Improving the sensitivity and specificity of detection technologies is essential for reliably recognizing gene fusions and alternative splicing events. The development of sequencing technologies, including long-read sequencing and single-molecule sequencing, has the potential to enhance the identification of intricate splicing patterns and rare fusion events (Tilgner et al., 2014).

Integration with multi-omics data

Combining transcriptome data with other omics data, including genomes, proteomics, and epigenomics, yields a more comprehensive understanding of cancer biology. Utilizing multi-omics techniques allows for the discovery of regulatory networks and pathways that are implicated in gene fusions and variations in splicing. An integrated study can reveal new treatment targets and biomarkers (Hasin et al., 2017).

Single-cell transcriptomics

Single-cell RNA-Seq is emerging as a powerful approach for investigating the variability in splicing at the individual cell level (**Figure 11.3**). This method offers valuable information on the range of gene expression patterns within tumors and identifies infrequent cell types that have a role in the development of diseases and resistance to therapy. Single-cell transcriptomics enables the detection of changes in the way genes are spliced in individual cells, providing a more comprehensive insight into their contribution to cancer (Lodato & Woodworth, 2019).

Therapeutic modulation of splicing

Directing interventions towards the splicing machinery and targeting individual splicing events has great potential as a therapeutic approach. Preliminary investigations have demonstrated the potential of small compounds and antisense oligonucleotides to manipulate splicing. Splice-switching oligonucleotides have been created to rectify abnormal splicing in illnesses including spinal muscular atrophy and Duchenne muscular dystrophy (Aartsma-Rus et al., 2016). Investigating the therapeutic manipulation of splicing in cancer is a current focus of study.

Extracellular RNAs as diagnostic indicators

Transcriptomics has greatly expanded our comprehension of gene expression and its implications in the context of health and disease. One of the growing fields of focus in transcriptomics is the investigation of extracellular RNAs (exRNAs). These RNA molecules, present extracellularly in several physiological fluids, have demonstrated significant potential as non-invasive biomarkers for a diverse array of disorders. Extracellular RNAs (exRNAs) have the capacity to revolutionize clinical diagnostics by serving as diagnostic markers,

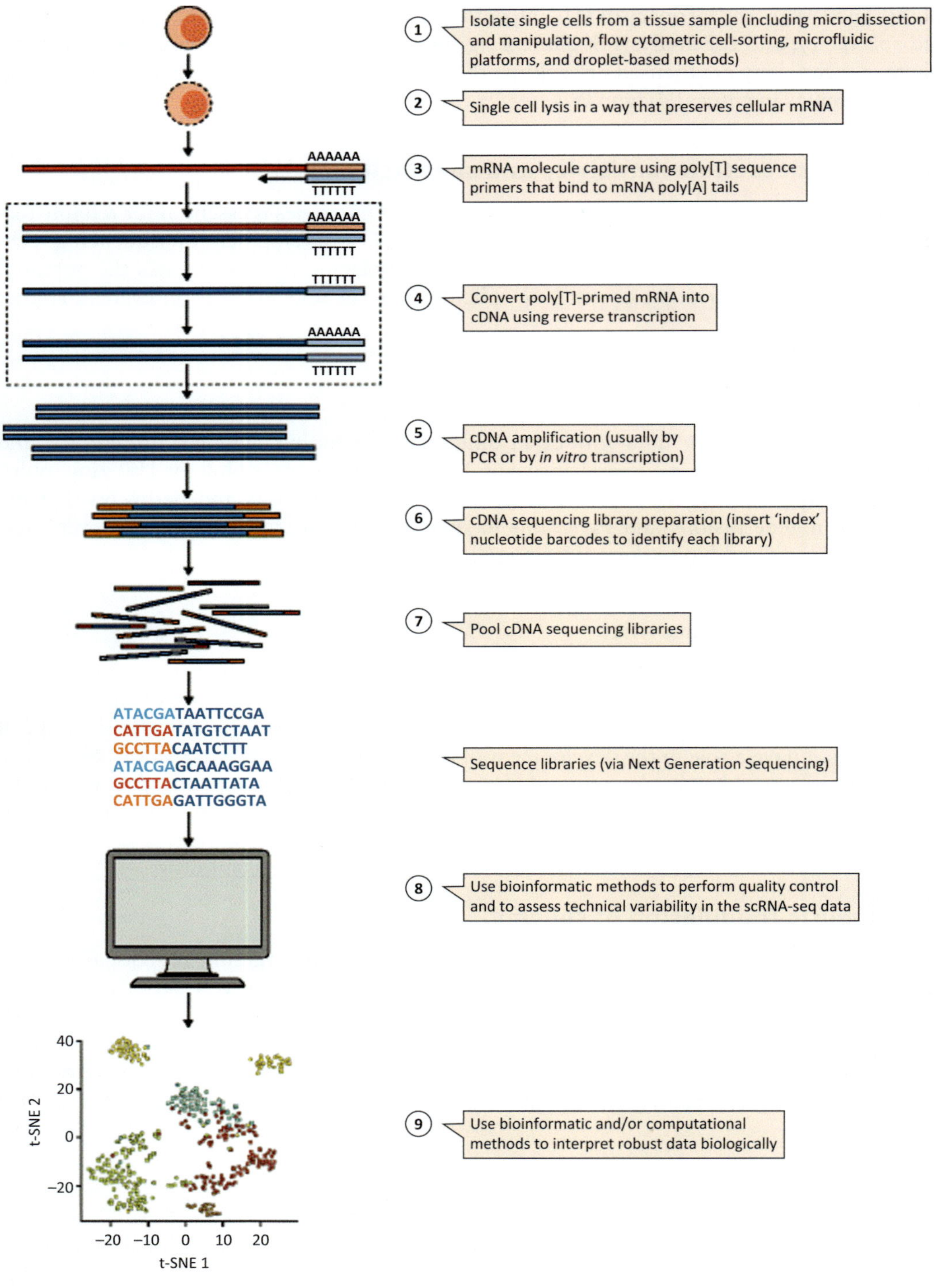

Figure 11.3 Single-cell RNA sequencing workflow illustrating cell isolation, mRNA capture, cDNA synthesis, library preparation, sequencing, and data analysis for single-cell gene expression profiling.

facilitating the early diagnosis and monitoring of diseases with little invasiveness. They are detected in different physiological fluids, including blood, urine, saliva, cerebrospinal fluid (CSF), and other biofluids. The exRNAs encompass a wide range of RNA varieties, each with unique biological functions and possible medical uses. The main categories of exRNAs are messenger RNAs (mRNAs), microRNAs (miRNAs), long non-coding RNAs (lncRNAs), and circular RNAs (circRNAs). The inclusion of these RNAs in extracellular vesicles (EVs), such exosomes and microvesicles, or their binding to proteins and lipoproteins, shields them from degradation by extracellular RNases and improves their stability in the extracellular surroundings.

TYPES OF EXTRACELLULAR RNAs

Messenger RNAs (mRNAs)

Messenger RNAs (mRNAs) are copies of genes that encode proteins and carry genetic instructions from DNA to the ribosome, where they direct the production of proteins. Regarding exRNAs, it is possible for mRNAs to be discharged into biological fluids either within EVs or attached to proteins. Despite being often found in lesser quantities than other forms of RNA, the identification and measurement of extracellular mRNAs can offer vital information on the gene expression patterns of particular cells or tissues. Biomarkers for a range of disorders, such as cancer, cardiovascular diseases, and infectious diseases, can be identified by the detection of certain mRNAs in biofluids.

MicroRNAs (miRNAs)

MicroRNAs (miRNAs) are short RNA molecules, around 20–25 nucleotides long, that have important functions in controlling gene expression after transcription. miRNAs form complementary base pairs with specific sequences on target mRNAs, resulting in either the destruction of the mRNAs or the prevention of their translation. Extracellular miRNAs are found in large quantities in body fluids and are frequently enclosed in EVs or linked to proteins such as Argonaute (AGO) and high-density lipoproteins (HDL). Due to their stability and ease of detection, they are highly desirable as non-invasive biomarkers. The abnormal expression of some miRNAs has been associated with a range of illnesses, such as cancer, cardiovascular diseases, neurological disorders, and metabolic problems.

Long non-coding RNAs (lncRNAs)

Long non-coding RNAs (lncRNAs) are RNA molecules that exceed 200 nucleotides in length and do not encode proteins. However, they play crucial roles in regulating gene expression. Long non-coding RNAs (lncRNAs) have a significant role in several biological activities, such as modifying the structure of chromatin, controlling gene expression, and regulating the processing of RNA molecules after transcription. Extracellular long non-coding RNAs (lncRNAs) are present in biofluids and have been associated with a range of illnesses. Specifically, some long non-coding RNAs (lncRNAs) exhibit distinct levels of expression in persons with cancer compared to those who are healthy, indicating their potential as biomarkers for the detection and prediction of cancer. Moreover, lncRNAs have been linked to cardiovascular ailments, neurological disorders, and inflammatory conditions.

Circular RNAs (circRNAs)

Circular RNAs (circRNAs) are a type of non-coding RNAs that have closed-loop structures, making them resistant to breakdown by exonucleases. Circular RNAs (circRNAs) are formed via a mechanism known as back-splicing, in which a splice donor site downstream is connected to a splice acceptor site upstream. These ribonucleic acids (RNAs) exhibit exceptional stability and may be identified in a range of biological fluids. Circular RNAs (circRNAs) have demonstrated their ability to function as miRNA decoys, effectively capturing miRNAs and inhibiting their interaction with their intended mRNA targets. The regulatory role and stability of circRNAs make them very attractive biomarkers for disease diagnosis and monitoring.

MECHANISMS OF RELEASE AND PROTECTION

Encapsulation within extracellular vesicles

EVs are particles enclosed by a membrane that cells discharge into the space outside of the cell. EVs encompass exosomes, microvesicles, and apoptotic bodies, which differ in terms of their size and source. Exosomes, which have a diameter ranging from 30 to 150 nm, are created inside multivesicular bodies and are released when they fuse with the plasma membrane. Microvesicles, which are bigger than exosomes, are formed directly from the plasma membrane by budding. EVs contain a variety of biomolecules, such as exRNAs, proteins, and lipids. These EVs provide a protective environment that safeguards exRNAs from being broken down by enzymes. The process of releasing EVs and their cargo is strictly controlled and impacted by factors such as cellular stress, hypoxia, and other stimuli.

Association with proteins and lipoproteins

Extracellular RNAs (exRNAs) can be found within EVs as well as coupled with proteins and lipoproteins in the extracellular environment. miRNAs frequently associate with AGO proteins, which are essential components of the RNA-induced silencing complex (RISC). This connection serves to safeguard miRNAs against degradation and enables their transportation to target cells. Likewise, some exRNAs are attached to HDL particles, which contribute to their durability and conveyance. The interactions between proteins and RNA, as well as lipoproteins and RNA, play a crucial role in maintaining the stability and functionality of extracellular RNAs (exRNAs) in biological fluids.

BIOLOGICAL FUNCTIONS AND INTERCELLULAR COMMUNICATION

Intercellular communication

Extracellular RNAs are essential for transmitting signals between cells in different physiological and pathological situations, hence facilitating intercellular communication. Recipient cells can internalize EVs harboring exRNAs by endocytosis, direct fusion with the plasma membrane, or receptor-mediated interactions. Upon entering the receiving cell, exRNAs have the ability to regulate gene expression and impact cellular activity. Tumor-derived EVs play a crucial role in cancer by facilitating tumor development, metastasis, and immune evasion. They achieve this by delivering oncogenic RNAs to neighboring cells.

Regulation of gene expression

Extracellular RNAs play an important role in controlling gene expression in the cells. For example, miRNAs transported by EVs might inhibit the expression of certain genes by attaching to complementary sequences on messenger RNAs (mRNAs). Likewise, long non-coding RNAs (lncRNAs) and circular RNAs (circRNAs) can influence the expression of genes in several ways, such as modifying the structure of chromatin, regulating transcription, and sequestering microRNAs (miRNAs). The regulatory roles of exRNAs are crucial for maintaining cellular balance and responding to external stimuli.

IMMUNE MODULATION

Extracellular RNAs (exRNAs) also have a role in regulating the immune response. EVs that contain microRNAs (miRNAs) have the ability to impact the function of immune cells by controlling the expression of genes that are connected to the immune system. Tumor-derived EVs can transport microRNAs (miRNAs) that inhibit the anti-tumor immune response, hence promoting immune evasion. In contrast, EVs secreted by immune cells have the ability to transmit microRNAs (miRNAs) and other regulatory RNAs to specific cells, so influencing their activity and playing a role in immune surveillance and inflammation.

Clinical significance of exRNAs

Detecting exRNAs in physiological fluids has important clinical consequences. Contrary to conventional biopsies that necessitate intrusive procedures, the examination of exRNAs provides a non-invasive method for diagnosing and monitoring diseases. This is especially beneficial for disorders in which acquiring tissue samples is difficult or hazardous.

Cancer diagnostics

Extracellular RNAs (exRNAs) are now being investigated as diagnostic biomarkers in the field of cancer research. Tumor cells secrete RNA molecules into the circulation, and these extracellular RNAs (exRNAs) can indicate the genetic and epigenetic changes that exist in the tumor. For instance, several miRNAs have been recognized as possible biomarkers for different types of malignancies. miR-21, miR-155, and miR-210 have been identified as elevated microRNAs (miRNAs) in the plasma of cancer patients, including those with breast, lung, and colorectal malignancies (Mitchell et al., 2008; Chen et al., 2008). Examining exRNAs can yield useful insights about the diversity of tumors, the effectiveness of treatments, and the advancement of diseases. Liquid biopsies, which examine exRNAs and other biomarkers in blood samples, are emerging as a potent tool for cancer diagnosis and tailored therapy. They provide the opportunity for early discovery, continuous monitoring of treatment effectiveness, and the detection of minimal residual disease and relapse.

Cardiovascular diseases

Extracellular RNAs (exRNAs) hold potential as diagnostic biomarkers for cardiovascular diseases (CVDs). Circulating miRNAs have been associated with numerous aspects of cardiovascular diseases (CVDs), such as myocardial infarction, heart failure, and atherosclerosis. For example, miR-1, miR-133a, and miR-208 are linked to damage in the heart muscle and may be identified in the bloodstream after a heart attack (D'Alessandra et al., 2010).

The non-invasive identification of these miRNAs enables early identification of cardiovascular events and the surveillance of disease progression. Furthermore, exRNAs have the potential to offer valuable information on the fundamental processes of cardiovascular diseases (CVDs), which might eventually result in the identification of new therapeutic targets.

Neurological disorders

Neurological disorders such as Alzheimer's disease, Parkinson's disease, and multiple sclerosis present significant difficulties in diagnosis. Extracellular RNAs (exRNAs) present in cerebrospinal fluid (CSF) and blood provide a non-invasive method for diagnosing and monitoring these disorders. For instance, some microRNAs (miRNAs) and long non-coding RNAs (lncRNAs) have been shown to exhibit differential expression in the CSF of individuals with Alzheimer's disease. This finding suggests that these molecules might serve as possible biomarkers for the early detection of the disease (Leidinger et al., 2013). Detecting exRNAs linked to neurodegenerative illnesses has the potential to enable earlier diagnosis and more effective therapy of these ailments, thereby enhancing patient outcomes and quality of life.

Infectious diseases

exRNAs can be used to diagnose and monitor infectious disorders, including viral and bacterial infections. Viral RNAs that are released into the circulation during an infection can be used as biomarkers to indicate the presence and severity of the disease. An example of an important diagnostic technique for viral diseases, including HIV, hepatitis B, and COVID-19, is the identification of viral RNA in blood samples (Wang et al., 2020).

Extracellular RNAs (exRNAs) can serve as indicators of the immunological response of the host to infections, offering valuable information about the development of diseases and the effectiveness of treatments. Due to its non-invasive nature, exRNA analysis is a significant tool for identifying and controlling infectious diseases, especially in situations with limited resources.

CHALLENGES AND ADVANCEMENTS IN THE ISOLATION AND QUANTITATIVE ANALYSIS OF EXTRACELLULAR RNAs

While the potential of exRNAs as diagnostic biomarkers is immense, several challenges must be addressed to fully realize their clinical utility. These challenges include the isolation of exRNAs, their stability in biofluids, and the quantitative analysis of their expression levels.

Quantitative analysis of extracellular RNAs

The quantitative analysis of exRNAs requires sensitive and accurate methods to measure their expression levels. Several techniques are employed for this purpose, each with its strengths and limitations.

Quantitative PCR (qPCR): Quantitative polymerase chain reaction (qPCR) is a commonly employed technique for measuring the abundance of certain extracellular RNAs (exRNAs). It provides a high level of sensitivity and specificity, enabling the identification of exRNAs that are present in low quantities. qPCR assays can be developed to selectively detect and measure the expression levels of certain microRNAs (miRNAs), messenger RNAs (mRNAs), or other types of RNA molecules. Nevertheless, the precision of qPCR can be influenced by variables like primer selection, reaction parameters, and sample integrity.

Digital PCR (dPCR): Digital PCR is an advanced form of PCR that allows for the absolute quantification of exRNAs. dPCR partitions the sample into thousands of individual reactions, enabling the precise counting of RNA molecules. This method is highly sensitive and can detect rare exRNAs with greater accuracy than qPCR. dPCR is particularly useful for detecting low-abundance biomarkers and for applications requiring high precision.

NGS: NGS technologies, including RNA sequencing (RNA-Seq), provide comprehensive profiling of exRNAs. NGS can identify and quantify a wide range of exRNAs, including miRNAs, mRNAs, lncRNAs, and circRNAs, in a single assay. This method offers high throughput and the ability to detect novel exRNAs. However, NGS requires sophisticated bioinformatics tools for data analysis and can be cost-prohibitive for routine clinical use.

Microarrays: Microarrays can be used to profile the expression levels of multiple exRNAs simultaneously. This method is suitable for high-throughput analysis and can provide valuable insights into the expression patterns of exRNAs. However, microarrays are limited by the need for predefined probes and may not detect novel exRNAs.

ADVANCEMENTS IN exRNA ANALYSIS

Recent advancements in exRNA analysis are addressing some of the challenges associated with their detection and quantification. These advancements include:

Improved Isolation Techniques: Novel isolation methods, such as size-exclusion chromatography, immunoprecipitation, and microfluidic-based approaches, are enhancing the purity and yield of exRNAs. These techniques offer greater specificity and efficiency in isolating exRNAs from biofluids.

Standardization and Quality Control: Efforts are underway to standardize protocols for exRNA isolation, processing, and analysis. The development of reference materials and quality control measures is essential to ensure reproducibility and reliability in exRNA research and clinical applications.

Integration with Other Biomarkers: Combining exRNA analysis with other biomarkers, such as proteins, metabolites, and genomic data, can provide a more comprehensive understanding of disease states. Multi-omics approaches are being explored to integrate exRNA data with other molecular information, enhancing the diagnostic and prognostic potential of exRNAs.

EMERGING TECHNOLOGIES IN TRANSCRIPTOMICS

Transcriptomics has made significant progress due to the emergence of cutting-edge technology that allows for more accurate, thorough, and efficient examination of RNA molecules. Emerging technologies have transformed the understanding of gene expression and its control, offering novel perspectives on cellular processes and disease mechanisms. The highly promising and emerging technologies in the field of transcriptomics, such as single-cell RNA sequencing, spatial transcriptomics, and third-generation sequencing, are discussed.

SINGLE-CELL RNA SEQUENCING (scRNA-Seq)

scRNA-Seq has transformed transcriptomics by allowing the examination of gene expression at the level of individual cells. scRNA-Seq, unlike bulk RNA sequencing, allows for the identification of variations in gene expression within a population of cells, providing insights into the unique transcriptional characteristics of individual cells. ScRNA-Seq has been greatly enhanced in terms of sensitivity, throughput, and accuracy due to recent technical advancements. Droplet-based techniques, such as Drop-seq and 10x Genomics Chromium, enable the concurrent analysis of several individual cells by enclosing them in nanoliter-sized droplets along with barcoded beads (Macosko et al., 2015; Zheng et al., 2017). These techniques facilitate the processing of large amounts of data and have been extensively utilized in a range of fields, such as developmental biology, cancer research, and immunology. scRNA-Seq has played a crucial role in finding rare cell types, elucidating cell development pathways, and revealing how cells respond to changes in their environment. For instance, scRNA-Seq has yielded valuable information on the tumor microenvironment, uncovering the existence of several subtypes of immune cells and their respective functions in tumor advancement and resistance to therapy (Tirosh et al., 2016). Furthermore, scRNA-Seq has been employed to delineate the developmental lineage of cells in the early embryo, providing insights into the molecular mechanics of embryogenesis (Cao et al., 2017).

SPATIAL TRANSCRIPTOMICS

Spatial transcriptomics is an emerging technology that integrates gene expression profiling with spatial information, enabling researchers to create a map of the transcriptome within the tissue environment. This approach overcomes a significant drawback of conventional transcriptomics, which usually lacks spatial resolution and cannot provide information on the spatial arrangement of gene expression.

Spatial transcriptomics technologies employ many methods for collecting gene expression data that is spatially resolved. An approach to do this involves the collection of mRNA from tissue slices using spatially barcoded arrays, which is then followed by RNA sequencing (Ståhl et al., 2016). An alternative method, referred to as in situ hybridization-based approaches, utilizes fluorescent probes to directly detect and measure RNA molecules inside tissue slices, resulting in detailed spatial maps of gene expression (Chen et al., 2015).

The field of spatial transcriptomics has significant consequences for understanding the arrangement of tissues, the interactions between cells, and the pathophysiology of diseases. The spatial transcriptomics in cancer research can uncover the geographic diversity of tumors, pinpointing specific cellular environments and their corresponding gene expression patterns. This knowledge is vital for understanding the process of tumor progression, metastasis, and the response to therapy. Spatial transcriptomics is a technique used in neuroscience to map the expression of genes in brain tissues, revealing the genetic basis of brain function and neurological diseases.

THIRD-GENERATION SEQUENCING

Third-generation sequencing methods, such as single-molecule real-time (SMRT) sequencing by Pacific Biosciences (PacBio) and nanopore sequencing by Oxford Nanopore methods (ONT), are a major advancement in the field of transcriptomics. These methods provide the

Table 11.1 Comparison of RNA sequencing platforms

Platform	*Principle*	*Read length*	*Throughput*	*Typical applications*
Illumina	Sequencing by synthesis	50–300 bp	High (up to terabases per run)	Gene expression profiling, alternative splicing
Pacific Biosciences	Single molecule real-timesequencing	>10,000 bp	Moderate (up to gigabases per run)	Full-length isoform sequencing, de novo assembly
Oxford Nanopore Technologies	Nanopore sequencing	Up to megabases	Variable (depends on flow cell used)	Direct RNA sequencing, long-read applications

real-time sequencing of lengthy RNA molecules, allowing for direct and precise understanding of the transcriptome (**Table 11.1**).

PacBio's SMRT sequencing and ONT nanopore sequencing provide several benefits compared to second-generation sequencing technology. The production of long reads is crucial for resolving intricate transcript structures, recognizing whole isoforms, and detecting RNA changes. Furthermore, these approaches eliminate the necessity for PCR amplification, hence mitigating amplification biases and conserving the original RNA sequence (Eid et al., 2009; Jain et al., 2016). Third-generation sequencing is highly advantageous for investigating alternative splicing, isoform diversity, and transcript variations. It offers a comprehensive perspective of the transcriptome, unveiling the whole sequences of RNA molecules and their changes. Precise comprehension of gene regulation and the functional outcome of alternative splicing heavily depends on this level of detail. In addition, third-generation sequencing may be used for single-cell transcriptomics, allowing for the examination of isoform variability at the individual cell level.

INTEGRATION WITH OTHER OMICS TECHNOLOGIES (PROTEOMICS, METABOLOMICS)

The combination of transcriptomics with other omics technologies, such as proteomics and metabolomics, is enhancing understanding of the intricate molecular networks that form the basis of biological processes and disease conditions. The utilization of multi-omics techniques offers an extensive perspective on cellular activity, facilitating the detection of molecular interactions and pathways that cannot be captured solely through transcriptomics.

Proteomics

Proteomics is the large-scale study of proteins, including their expression, modifications, interactions, and functions. While transcriptomics provides information about gene expression at the RNA level, proteomics offers insights into the final functional products of genes–proteins.

Integration with transcriptomics

The integration of transcriptomics and proteomics enables the identification of the relationship between mRNA and protein expression levels, therefore uncovering the mechanisms that regulate gene expression after transcription. Integrating this data can reveal differences in the levels of mRNA and protein, which can arise from several processes such as mRNA breakdown, control of translation, and protein breakdown (Vogel & Marcotte, 2012). Multi-omics investigations have the ability to uncover crucial regulatory proteins and the pathways they are related to, so offering a more thorough comprehension of cellular function and the mechanisms behind diseases. By combining transcriptomics with proteomics in cancer research, it is possible to find biomarkers that may be used for early identification, prognosis, and evaluation of therapy response. By integrating RNA-Seq data with mass spectrometry-based proteomics, it is possible to uncover the molecular characteristics of different forms of cancer, which may then be used to guide personalized treatment strategies (Zhang et al., 2014). The field of neuroscience may employ this integration to clarify the molecular foundation of neurological disorders by finding protein targets for therapeutic intervention.

Metabolomics

Metabolomics is the study of small molecules (metabolites) involved in metabolic pathways. Metabolites are the end products of cellular processes and reflect the physiological state of a cell or organism. Metabolomics provides a snapshot of cellular metabolism, offering insights into metabolic alterations associated with diseases.

Integration with transcriptomics

The integration of transcriptomics with metabolomics establishes a connection between gene expression and metabolic alterations, revealing the regulatory networks that govern cellular metabolism. This integration enables the identification of crucial metabolites and their corresponding pathways, therefore uncovering the functional implications of alterations in gene expression. Multi-omics methodologies can also reveal the intricate relationship between gene regulation and metabolism under varying circumstances.

By combining transcriptomics with metabolomics, metabolic illnesses like diabetes and obesity may be studied to uncover the precise molecular pathways that drive disease development and to pinpoint possible targets for therapeutic intervention. By integrating RNA-Seq with mass spectrometry-based metabolomics, researchers may detect metabolic signatures that are linked to insulin resistance and lipid metabolism (Su et al., 2014). By integrating several factors, such as the metabolic vulnerabilities of tumor cells, cancer research might identify targeted medicines that effectively disrupt cancer metabolism.

REAL-TIME AND RAPID TRANSCRIPTOMICS

The ability to perform real-time and rapid transcriptomics is transforming the landscape of molecular diagnostics, enabling timely decision-making in clinical and research settings. Advances in sequencing technologies and bioinformatics are driving the development of methods that provide near-instantaneous insights into gene expression.

REAL-TIME SEQUENCING TECHNOLOGIES

Nanopore sequencing is a type of real-time sequencing technology that allows for the direct and continuous study of RNA molecules as they go through a nanopore. This method offers prompt sequencing data, enabling quick identification and examination of alterations in gene expression. ONTs) have created portable sequencing devices, including the MinION, which enable the process of real-time RNA sequencing. These devices have the ability to generate lengthy sequences and identify changes in RNA molecules, allowing for real-time analysis of a wide range of genetic information (Jain et al., 2016). Nanopore sequencers are well-suited for point-of-care diagnostics and field-based research due to their mobility and speed. Real-time transcriptomics has significant applications in infectious disease diagnostics, enabling the rapid identification of pathogens and monitoring of disease outbreaks. For example, nanopore sequencing has been used to track the spread of viral infections, such as Ebola and COVID-19, providing critical information for public health responses (Quick et al., 2016; Wang et al., 2020). In cancer research, real-time transcriptomics can monitor treatment response, detecting dynamic changes in gene expression during therapy and guiding treatment adjustments.

RAPID TRANSCRIPTOME PROFILING

Rapid transcriptome profiling approaches are designed to expedite the process of generating and analyzing transcriptomic data, hence decreasing the time required from sample collection to obtaining useful insights. These approaches are crucial for applications that need rapid turnaround periods, such as clinical diagnostics and real-time monitoring of biological responses. Various methodologies have been devised to accomplish expeditious transcriptome profiling. Flash-frozen tissue samples, expedited RNA extraction processes, and optimized library preparation methods enhance the speed of data collection. In addition, advancements in bioinformatics pipelines allow for real-time data processing, resulting in nearly immediate outcomes.

CONCLUSION

Transcriptomics has significantly enhanced our understanding of gene expression and its relevance in the diagnosis and treatment of diseases (**Table 11.2**). Researchers may now examine gene expression with exceptional accuracy and thoroughness using advanced technologies such as high-throughput sequencing, microarray-based approaches, single-cell RNA sequencing, and spatial transcriptomics. These approaches have provided valuable

Table 11.2 Applications of transcriptomics in disease diagnosis

Application	*Methodolcgy*	*Benefits*	*Examples*
Cancer diagnostics	mRNA expression profiling, RNA-Seq	Identification of cancer subtypes, prognostic biomarkers	Breast cancer subtyping, detection of fusion genes in leukemia
Genetic disorder diagnosis	RNA-Seq, microarrays	Detection of aberrant splicing, novel transcript discovery	Identifying splicing mutations in muscular dystrophy
Infectious disease detection	RNA-Seq, targeted sequencing	Rapid identification of pathogens, monitoring of outbreaks	SARS-CoV-2 detection, HIV viral load monitoring
Neurological disorders	Single-cell RNA-Seq, spatial transcriptomics	Understanding cellular heterogeneity, mapping brain tissue	Single-cell analysis of Alzheimer's disease, Parkinson's disease research

information on mRNA expression patterns, gene fusions, and alternative splicing events. This information is crucial for the diagnosis and treatment of illnesses, especially malignancies. Integrating transcriptomics with other omics technologies, such as proteomics and metabolomics, offers a full understanding of cellular function and disease causes. Moreover, the introduction of real-time rapid transcriptomics is transforming clinical diagnostics, allowing for prompt and well-informed decision-making. As transcriptomics advances, its use in personalized medicine will become more prominent, providing new possibilities for early disease detection, treatment response monitoring, and targeted therapy development. This will ultimately enhance patient outcomes and revolutionize healthcare.

REFERENCES

Aartsma-Rus, A., Fokkema, I., Verschuuren, J., et al. (2016). The exon skipping approach to Duchenne muscular dystrophy: progress and challenges. Journal of Internal Medicine, 279(1), 52–61.

Bartel, F., Taubert, H., & Harris, L. C. (2002). Alternative and aberrant splicing of MDM2 mRNA in human cancer. Cancer Cell, 2(1), 9–15.

Bentley, D. R., Balasubramanian, S., Swerdlow, H. P., et al. (2008). Accurate whole human genome sequencing using reversible terminator chemistry. Nature, 456(7218), 53–59.

Brown, R. L., Reinke, L. M., Damerow, M. S., et al. (2001). CD44 splice isoform switching in human and mouse epithelium is essential for epithelial-mesenchymal transition and breast cancer progression. Journal of Clinical Investigation, 121(3), 1064–1074.

Cancer Genome Atlas Research Network. (2014). Comprehensive molecular profiling of lung adenocarcinoma. Nature, 511(7511), 543–550.

Cao, J., Spielmann, M., Qiu, X., et al. (2017). The single-cell transcriptional landscape of mammalian organogenesis. Nature, 566(7745), 496–502.

Chen, K. H., Boettiger, A. N., Moffitt, J. R., Wang, S., & Zhuang, X. (2015). Spatially resolved, highly multiplexed RNA profiling in single cells. Science, 348(6233), aaa6090.

Chen, X., Ba, Y., Ma, L., Cai, X., Yin, Y., Wang, K., ... & Zhang, C. (2008). Characterization of microRNAs in serum: a novel class of biomarkers for diagnosis of cancer and other diseases. Cell Research, 18(10), 997–1006.

Conesa, A., Madrigal, P., Tarazona, S., et al. (2016). A survey of best practices for RNA-Seq data analysis. Genome Biology, 17(1), 13.

Druker, B. J., Talpaz, M., Resta, D. J., et al. (2001). Efficacy and safety of a specific inhibitor of the BCR-ABL tyrosine kinase in chronic myeloid leukemia. New England Journal of Medicine, 344(14), 1031–1037.

D'Alessandra, Y., Devanna, P., Limana, F., Straino, S., Di Carlo, A., Brambilla, P. G., ... & Martelli, F. (2010). Circulating microRNAs are new and sensitive biomarkers of myocardial infarction. European Heart Journal, 31(22), 2765–2773.

Eid, J., Fehr, A., Gray, J., et al. (2009). Real-time DNA sequencing from single polymerase molecules. Science, 323(5910), 133–138.

Esteller, M. (2011). Non-coding RNAs in human disease. Nature Reviews Genetics, 12(12), 861–874.

Fang, Y., Fullwood, M. J., Chen, B. J., & Lim, B. (2018). RNA sequencing in cancer precision medicine. Genomics, Proteomics & Bioinformatics, 16(5), 207–215.

Guttman, M., Amit, I., Garber, M., et al. (2009). Chromatin signature reveals over a thousand highly conserved large non-coding RNAs in mammals. Nature, 458(7235), 223–227.

Hamadeh, H. K., Bushel, P. R., Jayadev, S., et al. (2002). Gene expression analysis reveals chemical-specific profiles. Toxicological Sciences, 67(2), 219–231.

Hasin, Y., Seldin, M., & Lusis, A. (2017). Multi-omics approaches to disease. Genome Biology, 18(1), 83.

Hodos, R. A., Kidd, B. A., Shameer, K., Readhead, B. P., & Dudley, J. T. (2016). In silico methods for drug repurposing and pharmacology. Wiley Interdisciplinary Reviews: Systems Biology and Medicine, 8(3), 186–210.

Jain, M., Olsen, H. E., Paten, B., & Akeson, M. (2016). The Oxford Nanopore MinION: delivery of nanopore sequencing to the genomics community. Genome Biology, 17(1), 239.

Katz, Y., Wang, E. T., Airoldi, E. M., & Burge, C. B. (2010). Analysis and design of RNA sequencing experiments for identifying isoform regulation. Nature Methods, 7(12), 1009–1015.

Kwak, E. L., Bang, Y. J., Camidge, D. R., et al. (2010). Anaplastic lymphoma kinase inhibition in non–small-cell lung cancer. New England Journal of Medicine, 363(18), 1693–1703.

Lee, Y., & Rio, D. C. (2015). Mechanisms and regulation of alternative pre-mRNA splicing. Annual Review of Biochemistry, 84, 291–323.

Leidinger, P., Backes, C., Deutscher, S., Schmitt, K., Müller, S., Frese, K., ... & Keller, A. (2013). A blood based 12-miRNA signature of Alzheimer disease patients. Genome Biology, 14(7), R78.

Lodato, M. A., & Woodworth, M. B. (2019). Single-cell transcriptomics and functional genomics in brain development. Nature Reviews Neuroscience, 20(1), 71–84.

Macosko, E. Z., Basu, A., Satija, R., et al. (2015). Highly parallel genome-wide expression profiling of individual cells using nanoliter droplets. Cell, 161(5), 1202–1214.

Mitchell, P. S., Parkin, R. K., Kroh, E. M., Fritz, B. R., Wyman, S. K., Pogosova-Agadjanyan, E. L., ... & Tewari, M. (2008). Circulating microRNAs as stable blood-based markers for cancer detection. Proceedings of the National Academy of Sciences, 105(30), 10513–10518.

Papalexi, E., & Satija, R. (2018). Single-cell RNA sequencing to explore immune cell heterogeneity. Nature Reviews Immunology, 18(1), 35–45.

Perou, C. M., Sørlie, T., Eisen, M. B., et al. (2000). Molecular portraits of human breast tumours. Nature, 406(6797), 747–752.

Quick, J., Loman, N. J., Duraffour, S., et al. (2016). Real-time, portable genome sequencing for Ebola surveillance. Nature, 530(7589), 228–232.

Roden, D. M., Pulley, J. M., Basford, M. A., et al. (2017). Development of a large-scale de-identified DNA biobank to enable personalized medicine. Clinical Pharmacology & Therapeutics, 92(1), 17–23.

Rowley, J. D. (1973). A new consistent chromosomal abnormality in chronic myelogenous leukemia identified by quinacrine fluorescence and Giemsa staining. Nature, 243(5405), 290–293.

Schwartz, R. E., Fleming, H. E., Khetani, S. R., & Bhatia, S. N. (2016). Pluripotent stem cell-derived hepatocyte-like cells. Biotechnology Advances, 34(5), 700–713.

Sibley, K., Cuthbert-Heavens, D., Knowles, M. A. (2001). Loss of heterozygosity at 4p16.3 and mutation of FGFR3 in transitional cell carcinoma. Oncogene. 20(6), 686–691. doi: 10.1038/sj.onc.1204110.

Soda, M., Choi, Y. L., Enomoto, M., et al. (2007). Identification of the transforming EML4-ALK fusion gene in non-small-cell lung cancer. Nature, 448(7153), 561–566.

Ståhl, P. L., Salmén, F., Vickovic, S., et al. (2016). Visualization and analysis of gene expression in tissue sections by spatial transcriptomics. Science, 353(6294), 78–82.

Su, G., Burant, C. F., & Beecher, C. W. (2014). Integrated metabolome and transcriptome analysis of the NCI60 dataset. BMC Bioinformatics, 12(1), 36.

Tilgner, H., Grubert, F., Sharon, D., & Snyder, M. P. (2014). Defining a personal, allele-specific, and single-molecule long-read transcriptome. Nature Biotechnology, 32(3), 284–289.

Tirosh, I., Venteicher, A. S., Hebert, C., et al. (2016). Single-cell RNA-seq supports a developmental hierarchy in human oligodendroglioma. Nature, 539(7628), 309–313.

Tomlins, S. A., Rhodes, D. R., Perner, S., et al. (2005). Recurrent fusion of TMPRSS2 and ETS transcription factor genes in prostate cancer. Science, 310(5748), 644–648.

Vogel, C., & Marcotte, E. M. (2012). Insights into the regulation of protein abundance from proteomic and transcriptomic analyses. Nature Reviews Genetics, 13(4), 227–232.

Wang, D., Johnston, S. L., & Meng, H. (2011). Pharmacogenomics in the era of personalized medicine. Clinical and Translational Science, 4(2), 119–120.

Wang, W., Xu, Y., Gao, R., Lu, R., Han, K., Wu, G., ... & Tan, W. (2020). Detection of SARS-CoV-2 in different types of clinical specimens. JAMA, 323(18), 1843–1844.

Wang, Z., Gerstein, M., & Snyder, M. (2014). RNA-Seq: a revolutionary tool for transcriptomics. Nature Reviews Genetics, 10(1), 57–63.

Wang, Z., Gerstein, M., Snyder, M. (2009). RNA-Seq: A revolutionary tool for transcriptomics. Nature Reviews Genetics *10*(1), 57–63. https://doi.org/10.1038/nrg2484

Zhang, B., Wang, J., Wang, X., et al. (2014). Proteogenomic characterization of human colon and rectal cancer. Nature, 513(7518), 382–387.

Zheng, G. X., Terry, J. M., Belgrader, P., et al. (2017). Massively parallel digital transcriptional profiling of single cells. Nature Communications, 8(1), 14049.

van 't Veer, L. J., Dai, H., van de Vijver, M. J., et al. (2002). Gene expression profiling predicts clinical outcome of breast cancer. Nature, 415(6871), 530–536.

Index

N

O

P

Q

R

S

T

U

V

W